335 Horsepower Flathead Ford V-8 Performance Handbook

by

Joseph P. Abbin, P.E.

Roadrunner Engineering
Albuquerque, New Mexico
505-268-6768
www.roadrunnerengineering.com

Layout, design and editing by Kathleen McCaughey.
Roadrunner logo by Corey Weintraub.
Front cover photo by Fouch Photography.
Most pictures by the author unless otherwise attributed.

Published by
Motorhead Mart
P.O. Box 53296
Albuquerque, New Mexico 87153
505-268-6768

Printed in the United States of America by
Starline Printing
7111 Pan American Highway NE
Albuquerque, NM 87109
800-873-7827

First Edition

ISBN – 13 978-0-615-26384-7

ISBN – 10 0615263844

Foreword

What! Another book on flatheads! Even though the "golden age" for the flathead has been over for 50 years or more, the flathead has refused to become extinct and is still providing plenty of street performance and "eye candy" for thousands of owners, many born after the last one was produced in 1953. Ford's "underhead cam" engine continues to fascinate generations of street rodders who can't resist the looks, the sounds and yes, the performance! The second "golden age" of the flathead may be now!

So what's new? How would we go about building a flathead today? That's what this book is all about. We will survey what was done "in the day" and evaluate the current options. Although years have passed since the flathead's heyday and technology has moved on, today's perspective validates most of the early rodder's work. For the most part, what worked then still works now. The purpose of this book is to provide proven guidelines for building a high performance flathead Ford for the street. Both normally aspirated and supercharged engines are covered. This book is not a detailed rebuilders manual. We will expand upon the basics covered in my previous book, ***Blown Flathead***. Key parts of that book have been reproduced in this book.

Since ***Blown Flathead*** was written, I have had the benefit of another 10 years of building, testing, street driving and racing flathead Fords. The acquisition of an engine dynamometer 4 years ago greatly accelerated the learning process and advanced the development of my computer modeling tools. I have also benefited greatly from the experience of my flathead supercharger kit customers, over 200 of them. In the same time period, the aftermarket has developed more new products for the flathead Ford than were developed in the previous 40 years! New and old speed equipment is now commonly available. Several new books on the flathead have been written and old books reprinted. Along with the new knowledge, a substantial number of flathead myths have also reappeared. The dyno and the street have helped separate fact from fiction. Not much has to be taken on faith today. The message is that flatheads make great street engines!

I doubt that the flathead fever has peaked, so consider this book a progress report. Stay tuned (pun intended)!

Blown Flatheads Forever,

Joe Abbin

Table of Contents

-1-

The Flathead Ford From Stock to Hot

Figure 1-1. Henry and Edsel Ford are shown admiring their favorite engine in this famous photo from the Ford archives. This is a first generation flathead as described in the text.

SUMMARY

In this chapter we will take a look at the evolution and characteristics of the 1932 to 1953 Ford and Mercury flathead V-8's. Following that, we'll take a look at developing a plan and a budget to improve street performance.

THE FLATHEAD FORD – AN EARLY PERFORMANCE ENGINE

The general pubic, including some infamous gangsters, immediately noticed the performance of the flathead V-8 after it was introduced in 1932. The early ads touted "75 mph in top gear!" By 1934, the top speed was up to 85 mph and the flathead Ford V-8 was established as the high performer in the low cost field, often superior in performance to autos costing much more. After some of the "bugs" in the first V-8's were sorted out, Fords dominated the street, as well as the track, until the mid-fifties. In addition to "stock" car racing of many types, the flathead even made it to the prestigious Indy 500, both before and after World War II. The terms "hot rod" and "flathead" were intimately linked during what many consider to be the "Golden Age of the Flathead," 1946-1952.

The flathead Ford did not disappear when Ford quit offering the engine in production cars over 50 years ago. Even though the modern OHV V-8's began appearing in 1949, the flathead remained common on the streets and the tracks for years afterwards. Flatheads continued racing successfully well into the fifties. The Kenz-Leslie streamliner held the American land speed record through 1958. Cal Rice's ***J.E. Riley Special*** finished off the mighty Chryslers to win the first NHRA Nationals in 1955. Even though Cal said goodbye to his flathead fuel dragster in 1955, he might have repeated his sucess at the 1956 Nationals if he had stuck with it. Both he and the 1956 National's winner, Mel Heath, failed to better the old flathead's ET that year with their Chryslers.

High performance flatheads continue to be raced in significant numbers even today. In 2002, I overheard one Bonneville official comment at Speed Week that "there are more flatheads running on the salt now than in 1952." If anything, that's still true today. Ron Main's ***Flatfire,*** the "World's Fastest Flathead," went over 304 mph there in 2004 and is currently the fastest Ford to date.

Even though interest in the flathead has had peaks and troughs since the "Golden Age," it has refused to become extinct and is still providing plenty of street performance and looks for thousands of owners, whether stock or modified.

THE STOCK FLATHEADS

In addition to the Ford and Mercury flathead V-8's discussed above, Ford also produced other flatheads during the 1932-1953 flathead era including the small V-8 60 (136 cu.in.), a large 337 cu.in. V-8, several V-12's, a six cylinder, and several four-cylinder engines. These engines will not be addressed in this book although some of the information presented is relevant. Unless otherwise specified, the term "flathead" in this book refers to engines based on the 221 to 255 cu.in. Ford and Mercury flathead engines produced in the US from 1932 to 1953.

FIRST GENERATION FORD FLATHEADS – 1932 to 1937

The flathead Ford V-8 family for purposes of this book consists of essentially three generations of engines with many variations along the way. I consider the first generation to be the engines built from 1932 to 1937. The first generation engines

Figure 1-2. Stock 1932 Ford Flathead. These engines displaced 221 cu.in. and were rated at 65 hp at 3600 rpm. Note the water pumps mounted on the heads, 21 head retaining studs and nuts, and a half bell housing built into the rear of the block. This engine is representative of the first generation engines described in the text.

are distinguished by their 21-stud head retainer pattern, the water pumps mounted on the front of the heads, the front mounted distributor, and the half bell housing built as part of the block. The flatheads produced in late 1937 had provision for the low mount water pumps used later.

The first generation flathead Ford V-8's hit the streets in March of 1932. This was Henry Ford's attempt to regain sales leadership from Chevrolet. Chevy had come out with a six cylinder engine in 1929, which many people considered to be quite a step up from the previous four cylinders in their own line and in Ford's Model A. Henry Ford felt that introducing a V-8 would leapfrog the competition and development was vigorously pursued to allow the engine to be offered in the 1932 lineup. The new V-8 ran great but had several issues including high oil consumption and a finicky 1-barrel Detroit Lubricator carburetor. It was also prone to overheating and vapor locking. By 1934, the engine was rated at 90 hp, primarily due to a new dual plane intake manifold with a Stromberg 2-barrel carburetor (Model 40), combined with an increase in compression ratio from 5.5 to 6.3:1. Oil consumption was brought under control by the addition of a second oil ring (four rings total on each piston). The second oil ring promoted better sealing and piston stability. Cooling was enhanced by the addition of a fourth impeller vane on the water pumps and a larger radiator. Vapor locking remained a flathead trademark due to placement of the fuel pump high in the back part of the engine. The 1934 Ford had the best power to weight ratio of any of the flathead era Fords, and better than most cars of any brand in 1934. This performance did not go un-noticed by the gangsters of the era (Ma Barker, Bonnie and Clyde, John Dillinger), or the early hot rodders. The relative scarcity and mechanical limitations of the first generation flatheads suggests that they should be saved for restored stock vehicles.

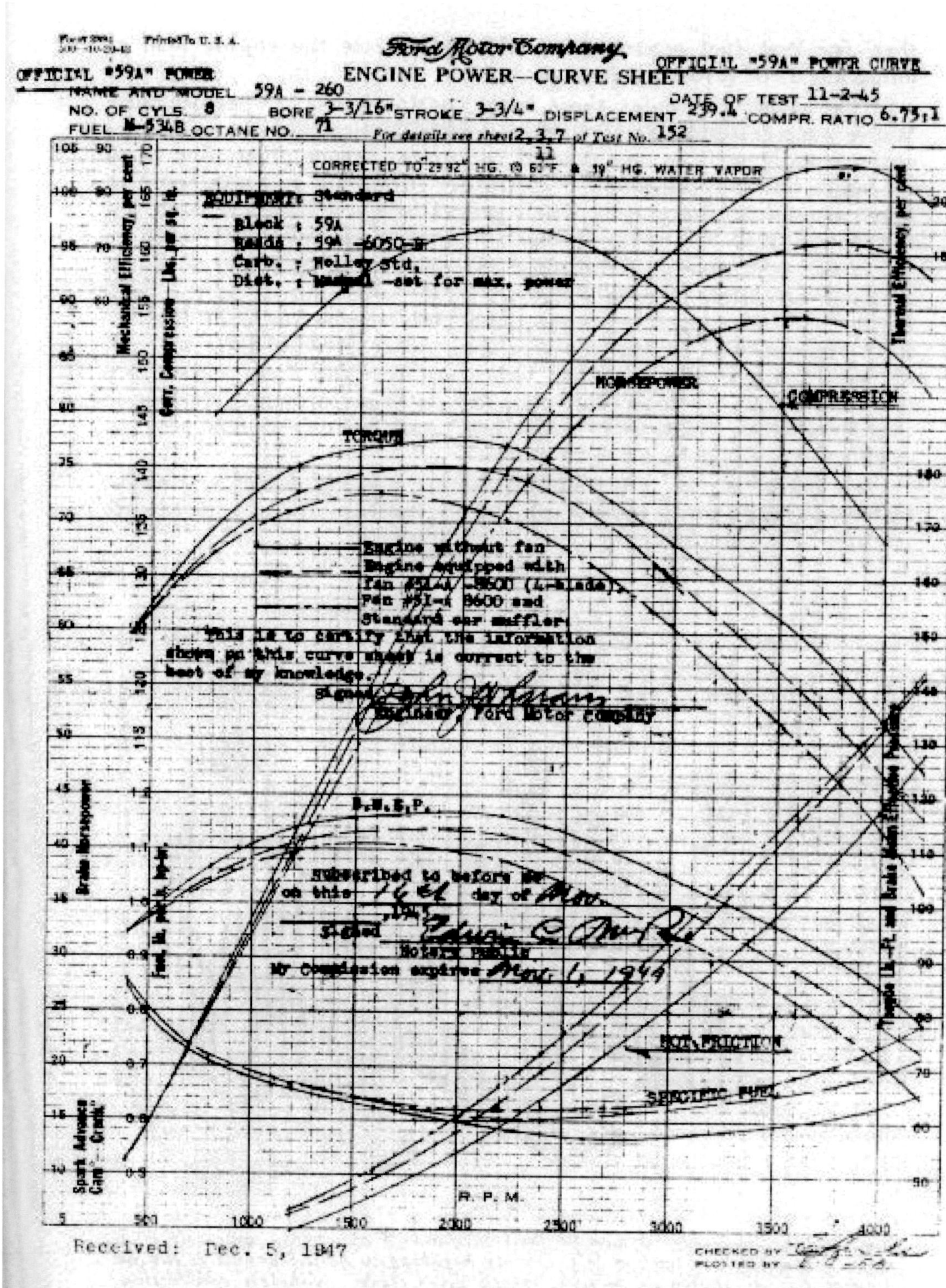

Fig. 2-6. *Performance sheet for the post-war Ford V8 engine.*

Figure 1-3. This factory dyno sheet from 1945 contains a large amount of data and makes a good basis for comparison with test results presented in later chapters. The three curves for horsepower with and without the fan and/or the muffler help explain the different ratings claimed in different sources for the same engine.

Year	Ford (F) Merc (M)	Model	Bore (inches)	Stroke (inches)	Disp (cu.in.)	Compression Ratio	Rated HP @ RPM	Torque (ft-lbs)	Comments
1938	F	81A	3-1/16	3-3/4	221	6.12	85@3800	146@2000	First 24 stud heads
1939	F	91A	3-1/16	3-3/4	221	6.15	90@3800	155@2200	
	M	99A	3-3/16	3-3/4	239	6.3	95@3800	170@2100	
1940	F	01A	3-1/16	3-3/4	221	6.15	90@3800		Same as 1939
	M	09A	3-3/16	3-3/4	239	6.3	95@3800		
1941	F	11A	3-1/16	3-3/4	221	6.15	90@3800		Same as 1939
	M	19A	3-3/16	3-3/4	239	6.3	95@3800		
1942	F	21A	3-1/16	3-3/4	221	6.2	96@3800	155@2000	
	M	19A	3-3/16	3-3/4	239	6.4	100@3800	180@2000	
1946	F	69A	3-3/16	3-3/4	239	6.75	100@3800	180@2000	
	M	69M	3-3/16	3-3/4	239	6.75	100@3800	180@2000	
1947	F	79A	3-3/16	3-3/4	239	6.75	100@3800		Same as 1946
	M	79M	3-3/16	3-3/4	239	6.75	100@3800		
1948	F	89A	3-3/16	3-3/4	239	6.75	100@3800		Same as 1946
	M	89M	3-3/16	3-3/4	239	6.75	100@3800		
1949	F	8BA	3-3/16	3-3/4	239	6.8	100@3600	180@2000	
	M	9CM	3-3/16	4	255	6.8	110@3600	200@2000	
1950	F	0BA	3-3/16	3-3/4	239	6.8	100@3600	181@2000	
	M	0M	3-3/16	4	255	6.8	110@3600	200@2000	
1951	F	1BA	3-3/16	3-3/4	239	6.8	100@3600	187@1800	
	M	1M	3-3/16	4	255	6.81	112@3600	206@2000	
1952	F	B2	3-3/16	3-3/4	239	7.2	110@3800	194@1900	
	M	BC	3-3/16	4	255	7.2	125@3700	218@1700	
1953	F	B3	3-3/16	3-3/4	239	7.2	110@3800	196@2000	
	M	BG	3-3/16	4	255	7.2	125@3800	218@1700	

Table 1-1. Partial Summary of Ford and Mercury Flathead V-8 Specifications.

SECOND GENERATION FORD FLATHEADS - 1938 to 1948

The second generation flatheads were built from 1938 to 1948. By 1938, the heads were retained by 24 studs, the water outlets from the block were in the top center of the heads, and the water pumps were located at the lower part of the front of the block. The front mount distributor and the half bell housing on the block remained. As the flathead evolved, cooling was further improved by higher capacity water pumps with six impellers, more effective radiators, pressurized cooling systems (1941), bigger fans and improved cooling water routing. Horsepower also improved over time with more displacement (239 cu.in. available in 1939 Mercury), less restrictive exhausts, increased compression ratio, and more aggressive camshafts. The rated horsepower fluctuated depending on what accessories were installed during testing and the sales emphasis. See factory dynamometer curve (Figure 1-3). During the 1946-1952 "golden age of the flathead," the second generation flatheads, particularly the 1946-1948 Fords and Mercs, became the hot rodders' favorites because they were widely available, both new and used. Speed equipment was also widely available, and the 1948 and earlier engines could be easily stroked with the drop-in addition of the 1949 Mercury crankshafts available from the dealer. The second generation engines, including the venerable "***59A***" blocks (1946-1948 Fords and Mercs), are rapidly approaching the scarce status.

Figure 1-4. Stock 1939-1941 Ford Flathead. These displaced 221 cu.in. and were rated at 85 hp at 3600 rpm. The Mercury equivalent displaced 239 cu.in. and was rated at 95 horsepower. Note the water outlets at the center of the head and 24 head retaining studs and nuts. The half bell housing is still part of the block. This engine is representative of the second generation engines described in the text.

THIRD GENERATION FORD FLATHEADS - 1949 to 1953

The third generation went from 1949 to 1953 in the US. See Figures 1-5 through 1-8. The last-generation flatheads are most easily recognized by the absence of the half bell housing on the block, and the water outlets at the top front of the heads. The last generation flatheads are also distinguished by the more conventional looking high mount distributors. Cooling was further improved by adding yet another two impellers to the water pumps and substantially redesigning the flow passages in the block and the heads. See Appendix B. The horsepower continued to increase in both the Fords and Mercurys due to higher compression ratios and more aggressive camshaft timing. The Mercurys of this generation were the horsepower kings with the ¼" longer stroke resulting in 255 cu.in., and larger capacity carburetors yielding 125 horsepower in 1953.

Some of the changes in power, displacement, and compression ratio of the second and third generation flatheads are summarized in Table 1-1 reprinted from ***Blown Flathead***.

Because of the many improvements and current availability, I recommend the third generation flatheads as the basis for a high performance street engine. Thus the emphasis in this book is on the third generation flatheads, referred to as the "***8BA***" style Fords and Mercurys.

DOING OUR HOMEWORK – BUY THE BOOK FIRST

One of my favorite sayings is "buy the book before the blower." That really goes for almost anything, including blowers. Doing the homework and planning are fundamental to a successful project. There are a number of references and suppliers listed in ***Blown Flathead*** and elsewhere to help the novice get started on building a flathead. I recommend reading everything you can find. Since ***Blown Flathead*** and its references were written, there has been much more written and published on the subject of building a flathead Ford. Readers will quickly note that the experts don't always agree, and are often contradictory. Thus the consumer must ferret out the truth, taking into account the content of the material, the credentials of the authors, and most importantly, **the actual demonstrated results**, not just the projections. Read carefully, because it is often hard to separate fact from fantasy! Be skeptical!

Realistically, my own dyno testing has demonstrated that a **well-built**, normally aspirated street flathead will produce 150-220 hp. A mild supercharged flathead will produce 225-275 hp (or more) and still run on premium pump gas. See Chapters 7 through 10. While these numbers will not impress modern engine ad readers, doubling and even tripling the stock flathead horsepower in a light car will produce a very impressive and distinctive street machine. See Chapter 2.

COST

The cost of building a flathead depends on several important factors including what you are starting with (do you have cores for the block, crank, rods, etc.), your performance goals (speed costs), your appearance goals (shinier costs more), and who does the work. A normally aspirated, **professionally built**, mildly modified flathead will cost about $4000 to $5000, if you have a re -buildable core. Add a satin finish blower kit and you're up to $7500. An example of such an engine is the Downs engine in Chapter 7. A start-from-scratch, polished, supercharged performance engine, such as the Honest Charley-Coker Tire engine in Chapter 8, can cost twice as much including the purchase price of cores. These costs are in 2006 dollars. Not cheap! One of my customers made the remark that "never had he spent so much for so little horsepower." The number of people either keeping their flathead or replacing a previously

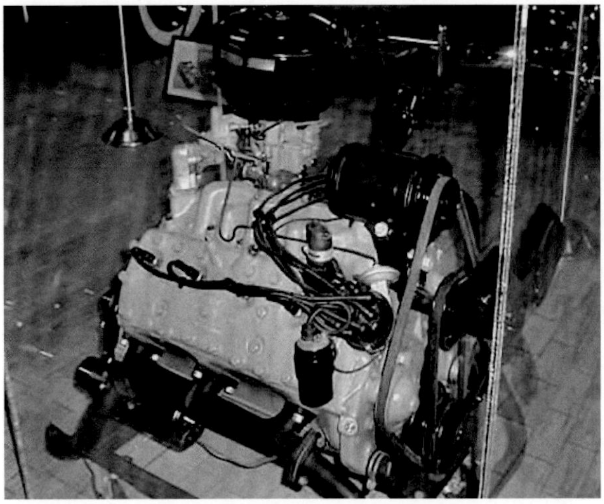

Figure 1-5. Stock 1953 Ford Truck Flathead, Passenger Side. These engines displaced 239 cu.in. and were rated at 110 hp at 3600 rpm. Note the top mounted conventional distributor. This engine is representative of the third generation engines described in the text.

Figure 1-6. Stock 1953 Ford Truck Flathead, Driver's Side. Note water outlets are now at the front of the heads like the first generation flatheads, but the water pumps are mounted on the block like the second generation flatheads.

Figure 1-7. Stock 1949-1951 Mercury Passenger Car Flathead. These engines displaced 255 cu.in. and were rated at 110 hp. The half-bell housing is bolted to the back of the block like the 1948–1952 Ford truck engine in Figures 1-5 and 1-6. This is another characteristic of the third generation engines.

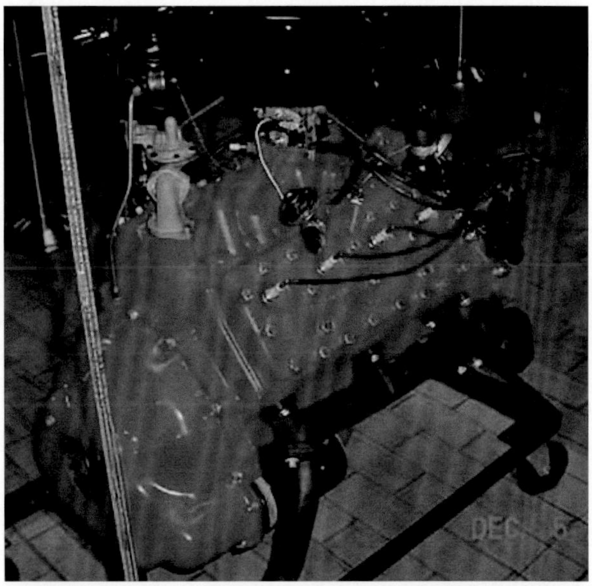

Figure 1-8. Stock 1949-1953 Ford Passenger Car Engine. These engines displaced 239 cu.in. and were rated at 100-110 hp. A full bell housing is bolted to the back of the block. Mercury adopted this bell housing in 1952-1953. These are also third generation engines.

swapped overhead indicates that the expense is worth it. Just say "no" to crate engines.

PLANNING AND BUDGETING

The fundamental questions in this phase of our project become: what do I want, where can I get it, and how much does it cost? Once we know what we want, and have decided that it's possible and affordable, the next step is a basic plan and budget.

Almost everything you could possibly need for building a flathead is available, but delivery times are often longer and prices are usually more expensive than equivalent parts and services for a common overhead valve engine. It certainly pays to shop around. A factor of two over delivery times and prices for equivalent overhead parts and services is not a bad starting point. However, just think about all the money you'll save on rocker arms and push rods!

At least as important as time and dollars however, is the choice of parts. Any successful engine build is dependent on the proper combination of parts. Obviously not everything goes together, even though we may want it to, or because we own it, or it's free. An example of concepts that don't go together is supercharging and high compression ratio. You might just have to make clocks out of those old high compression heads! An experienced engine builder can be a major asset in your buildup as long as they are cognizant of the differences between building and tuning a flathead vs. an overhead valve engine.

Chapter 2 will describe the ***Motorhead Mart Special***, certainly one of the quickest flathead powered street rods. Chapters 3, 4, 5 and 6 provide general discussions of major engine components and the factors involved in choosing a combination.

Without becoming a textbook, I believe this book can most effectively aid the performance flathead engine builder through the use of case studies of actual engine builds, i.e. complete engines, including tuneups. Chapters 7, 8, 9 and 10 provide descriptions of actual engines and their measured performance.

The discussions, the case studies, and the test results will give you an idea of what is available, what works, and what you can expect.

Chapter 10 in particular, describes the current engine in the ***Motorhead Mart Special***, nicknamed "***Killer.***" ***Killer*** is certainly one of the highest performing flathead street engines that I am aware of. **This is not to suggest that a more powerful flathead street engine can't be built or hasn't already been built.** However, ***Killer*** is very well documented all the way from detailed flow testing and engine dyno testing, through installation in a vehicle, and track testing. We don't race flow benches or dynos. However, I do drive and race the ***Motorhead Mart Special*** and the asphalt never lies! The ***Special*** is available for match races!

The ***Motorhead Mart Special*** and ***Killer*** demonstrate that a very strong flathead-powered street rod can be built with commonly available parts at a reasonable cost. They provide a basis from which we can pick and choose proven technology.

Let's get started!

A PREVIEW OF MAJOR POINTS IN THIS BOOK

- A well-built, normally aspirated street flathead will produce 150-220 hp. A mild supercharged flathead will produce 225-275 hp (or more). Pages 82, 88, 96, & 110.
- Blocks with cracks in the valve pocket area or center main cap web area are not recommended for high performance applications. Page 21.
- For proper cooling, it is extremely important to start with a clean block. Page 117.
- Do not use the 1948 and earlier heads (or head gaskets) on 1949 and up engines. Cooling is compromised. Page 119.
- Thread sealer should be used for all head bolt and stud installations to prevent coolant "weeping" in service. Page 21.
- Properly torqued deck plates and main caps should be used during cylinder boring and honing operations to enhance cylinder roundness and ring seal. Page 21.
- Heavy-duty main caps, "straps," etc. should be used to strengthen the flathead crankshaft support on high performance engines. Usually only the center main is reinforced. Mild street engines do not require any support. Page 21.
- Flat top pistons should not be used for any application without a matching flat combustion chamber head configuration. Page 22.
- No matter what heads are used, testing has clearly demonstrated that a traditional block "relief" substantially improves flow and power. Page 28.
- The optimum compression ratio for drivability and peak horsepower for a normally aspirated street flathead is about 8:1 to 9:1. Page 37.
- Peak power for most normally aspirated flatheads is obtained with total ignition advance numbers of 22-26 degrees. Page 55.
- Pre-ignition and detonation are serious threats to any engine, especially supercharged engines. Page 64.
- For supercharging, we must first determine our compression ratio. Page 35. Then we can pick a safe boost level for the fuel octane we choose. Page 65. Finally, we can determine a supercharger pulley ratio to provide the desired boost. Page 68.
- Roots-type blower capacities of about one-half to two-thirds the engine displacement per revolution work very well for the street. Page 67.
- Supercharged boost levels of 5 to 6 psi combined with 7.2:1 to 8.2:1 compression ratios are practical combinations for everyday driving at sea level. Page 66.
- Peak power for most supercharged flatheads can be obtained with total ignition advance numbers of 16-20 degrees. Page 71.
- Ignition boost retard modules can provide automatic electronic boost retard for any battery ignition, either points or electronic. These modules are recommended for all supercharged engines. Page 71.
- A positive crankcase ventilation (PCV) system venting to the base of the carburetor on a supercharged engine can initiate detonation when boost and blow-by are maximum. Page 122.
- After initial head installation, at least four warm up, cool down, and re-torque cycles are required for a good head-to-block seal. Page 124.
- SAE 20W50 racing oil meets the requirements of most flathead Fords. Page 126.

-2-

The Most Powerful Flathead Ford Powered Street Rod Ever?

The Motorhead Mart Special

Figure 2-1. The *Motorhead Mart Special* at the 1997 Goodguys March Meet in Bakersfield, CA. At the same track, the car ran a 12.41 second elapsed time and 107.6 mph at the 2006 NHRA California Hot Rod Reunion.

SUMMARY

In this chapter we will take a look at the ***Motorhead Mart Special***, perhaps the most powerful flathead powered street rod ever built. Certainly, it is one of the best documented flathead powered cars. The ***Special*** has served as a test bed for my company, Roadrunner Engineering, for approximately 14 years. During that time it has had two engines, three superchargers, and many different tuneups and driveline modifications. Both engines have been thoroughly documented from initial choice of components, hardware modifications, flow testing of the block assembly, and extensive track testing (over 300 drag strip runs) of the engines in the vehicle. In addition, the current engine has seen extensive dynamometer testing of the completed engine both normally aspirated and supercharged. See Chapter 10 for details on the engine, nicknamed ***Killer***.

Figure 2-2. The current engine in the *Motorhead Mart Special*. This engine has been dyno tested at over 335 hp and 365 ft-lbs. The engine is described in Chapter 10.

THE FLATHEAD AS A MODERN PERFORMANCE ENGINE?

Most people have noted the recent resurgence of flathead Ford powered vehicles at car shows and auto events across the country. Most of this resurgence results from a rediscovery of the timeless good looks and sound of the classic flathead Ford. However, even diehard flathead Ford fans concede that the flathead is obsolete as a performance engine by today's standards. But is it? **The *Motorhead Mart Special* is a street-driven, 1934 Ford Tudor sedan with a 276 cu.in. flathead that accelerates from 0-60 mph in 3.5 seconds!** Compare this with 4.3 seconds for a Viper SRT10 Coupe or 4.1 seconds for a 2006 Z06 Corvette (per Automobile Magazine). Who says the flathead should no longer be considered a performance engine? Even most skeptics agree that the ***Motorhead Mart Special*** is a flathead powered car that runs with the big dogs. You can't beat good looks, great sound, and performance too! Not everybody aspires to this level of performance of course, but we can all benefit from what has been learned, and take from it what we need and can afford.

Besides the engine in the ***Special***, the following chapters will provide general information on several other engine configurations and provide details helpful for any flathead buildup.

HISTORY OF THE *MOTORHEAD MART SPECIAL*

In 1991, I acquired a "project" 1934 Ford sedan. The car had sat in a barn for approximately 30 years after being campaigned as a drag car in the Midwest. The car had been crudely chopped and channeled with welded-on, bobbed fenders. What a mess!

My friends and I completely rebuilt the car from the ground up and put it back on the

Figure 2-3. The original *Motorhead Mart Special* engine in 1998. This engine featured a 9" Camden supercharger, Mercury *8CM* heads, and was mildly ported but not relieved. The engine is described in more detail in *Blown Flathead*. It produced approximately 280 hp in race trim and powered the *Special* to a best quarter mile performance of 100.4 mph in 13.41 seconds. It lives on in my Mercury. See Figure 2-4.

Figure 2-4. The original engine from the *Motorhead Mart Special* in its new home in the author's 1950 Mercury. This engine has required only routine service after 300 drag strip passes, 5000+ street miles, and 14+ years service. It moves the 3400 lb vehicle to 60 mph in 7 seconds and gives 19 mpg on the highway!

street in 1994 with a '49 Mercury flathead utilizing the prototype Roadrunner Engineering flathead supercharger kit. See Figure 2-3. In that configuration the car ran at strips in NM, AZ, CO, and CA and showed at numerous events, garnering many trophies and awards. Also during that time, the ***Motorhead Mart Special***, aka the "Pumpkin" (because of its orange paint), accumulated almost 2000 street miles with no problems. A high point in this period was at the 1998 Goodguys' March Meet in Bakersfield, CA where the car turned 100 mph in the quarter mile for the first time. The experience gained in building and running the ***Special*** was documented in ***Blown Flathead***, which provides the basics for building a high performance flathead both normally aspirated and supercharged. The book describes the engine and the car as it existed in 2000.

In 2003, the engine was removed from the ***Special*** and placed in my 1950 Mercury (See Figure 2-4) where it has accumulated another 3000+ miles, and continues to provide exceptional performance on the street.

THE NEXT GENERATION: DEVELOPMENT OF THE "KILLER" FLATHEAD

As good as the ***Special*** ran, and as reliable as it was, several areas for improvement were identified. I decided to build a new engine utilizing gradually accumulated knowledge from my own testing and computer modeling, and the operating experience of over two hundred Roadrunner customers.

The new engine was completed and began dynamometer testing in December 2005. After approximately two months of testing and approximately 80 "pulls" on the dyno, the results were in. The new engine, nicknamed, ***Killer***, demonstrated over 220 hp normally aspirated with three 2-barrel carburetors. Supercharged, it demonstrated over 365 ft-lbs of torque @ 4000 rpm and 335 hp @ 5000 rpm! Test rpm was limited to save wear and tear on the engine, so these numbers may not represent maximum potential of the engine. Nevertheless, they may represent the highest documented numbers for a street driven flathead on gasoline. See Chapter 10.

In May 2006, the ***Killer*** was installed in the ***Motorhead Mart Special***. In June, the ***Special*** ran at the High Altitude Flathead and Inline Nationals in Denver (track elevation 5800 feet). There the ***Special*** turned a best of 101 mph with a 13.3 second ET in the quarter mile. After winning the class final, the ***Special*** was awarded the 2006 "Top Rod" eliminator title.

Following the win at Denver, the ***Special*** was invited to be an exhibition car at the NHRA Hot Rod Reunion in Bakersfield, CA. During track runs at the Famoso Drag strip on October 4 and 5, 2006, the ***Special*** did not disappoint with all runs less than mid-12 seconds. The weekend's bests included a top speed of 107.6 mph and a quickest elapsed time of 12.41 seconds in the quarter mile. Again, these may be the quickest documented times for a street flathead. In contrast with Denver, Bakersfield's elevation is less than 400 feet. Even supercharged cars like low altitude!

See Table 2-1 for a complete summary of the ***Special's*** best drag strip performances.

MOTORHEAD MART SPECIAL SPECIFICATIONS

Vehicle
- 1934 Ford 2 door sedan. Chopped 3 inches. All steel.
- Stock frame.
- An old dropped axle with owner made hairpins. F-100 steering box.
- Weight 2540 lbs with driver (2740 lbs with fenders, running boards and street tires).

Engine (See Chapter 10)
- 1951 Mercury flathead, 276 cu.in.
- Race prepped stock rods and crankshaft. Balanced rotating assembly.
- Ross forged pistons with mm ring pack.
- Block ported and relieved.
- Modified Offenhauser ***425*** heads.
- Iskenderian ***400Jr*** camshaft and ***185G*** single valve springs.
- Manley 1.6" intake and exhaust valves.
- Roadrunner Engineering ***Big Kahuna*** street supercharger kit.
- Weiand ***174*** supercharger.
- Holley 600 cfm double pumper 4-bbl carburetor.
- Mallory dual point distributor and coil with MSD ***6BTM*** box.

Driveline
- Ford C4 automatic transmission with Flat-O adapter and 10" converter.
- Ford 9" rear end with Detroit Locker and 4.11 gears.
- Pete & Jake's ladder bars with stock rear spring.

Wheels and Tires
- Front: 165R15 on 1946 Mercury rims or 26x4.5-15 M/T tires on Halibrand mags.
- Rear: P255/70R15 on steel truck rims or 26x9-15 M/T slicks on Halibrand mags.

WHAT NOW?

The ***Motorhead Mart Special*** has always been a dual-purpose car, capable of cruising the street as well as competing at the track. To further demonstrate this dual capability, a full set of fenders, running boards, etc. were installed on the ***Special*** in February 2007. The car in this configuration made exhibition runs at the Standard 1320 Club meet at Speedworld in Phoenix on April 20, 2007. There the car, about 200 lbs heavier than at Bakersfield, still blasted the drag strip with a top speed of 103 mph and an elapsed time of 12.8 seconds. The altitude there is about 1200 feet. Movies of all these events are available, as well as the time slips, from Roadrunner Engineering.

The following chapters will focus on helping others reach or exceed these levels of performance with their street driven flatheads. In Chapter 10 you will find the "secrets" of the ***Killer*** engine including buildup and tuning. As you will see there is nothing exotic, just the details. As an inventor once noted, "the answers once known, are obvious." The information presented will be of value to racers as well as those who just want a reliable cruiser that looks, sounds and runs good.

Figure 2-5. The *Motorhead Mart Special* at Denver after winning "*Top Rod*" at the High Altitude Flathead-Inline Nationals in 2006. Shown left to right are the author, Kent Hendershot, and Jeff Andreen. Kent and Jeff performed a major part of the *Killer*'s engine build. This was the third *Top Rod* title for the *Special*.

Figure 2-6. The *Motorhead Mart Special* with full street equipment at the 2007 Sandia Rod & Custom Reunion in Albuquerque, NM. *Bill Varela pho-*

Figure 2-7. The *Motorhead Mart Special* enjoying a photo shoot in the foothills of the Sandia Mountains, Albuquerque, New Mexico.

Date	Time	60' ET	1/8 Mile ET Seconds	1/8 Mile TS MPH	1/4 Mile ET Seconds	1/4 Mile TS MPH	Temp Deg F	Baro Inches Hg	Humidity %
Flathead/Inline Nationals, Bandimere Speedway, Morrison, CO. Elevation 5800 Feet. Vehicle Weight 2540 lbs.									
6/3/2006	12:39	1.85	8.45	81.07	13.33	100.20	96	24.2	10
6/3/2006	14:51	1.91	8.55	80.29	13.46	100.13	88	24.2	10
6/4/2006	9:43	1.84	8.41	80.89	13.28	100.69	78	24.3	35
NHRA Hot Rod Reunion, Famoso Raceway, Bakersfield, CA. Elevation 400 Feet. Vehicle Weight 2540 lbs.									
10/6/2006	10:58	1.73	7.88	86.09	12.44	107.60	68	29.4	55
10/6/2006	16:40	1.72	7.90	85.68	12.49	106.71	81	29.3	36
10/7/2006	9:38	1.69	7.85	86.59	12.41	107.30	74	29.4	49
Standard 1320 Club, Speedworld Dragstrip, Phoenix, AZ. Elevation 1200 Feet. Vehicle Weight 2740 lbs (fenders, etc.).									
4/14/2007	10:00	1.74	8.07	82.39	12.84	102.45	81	28.5	28
4/14/2007	12:24	1.72	8.14	82.23	12.94	101.84	88	28.5	24
4/14/2007	14:21	1.75	8.12	82.32	12.89	102.54	94	28.4	15

Table 2-1. *Motorhead Mart Special* Best Drag Strip Performance Summary

-3-

Short Block Component Selection and Preparation

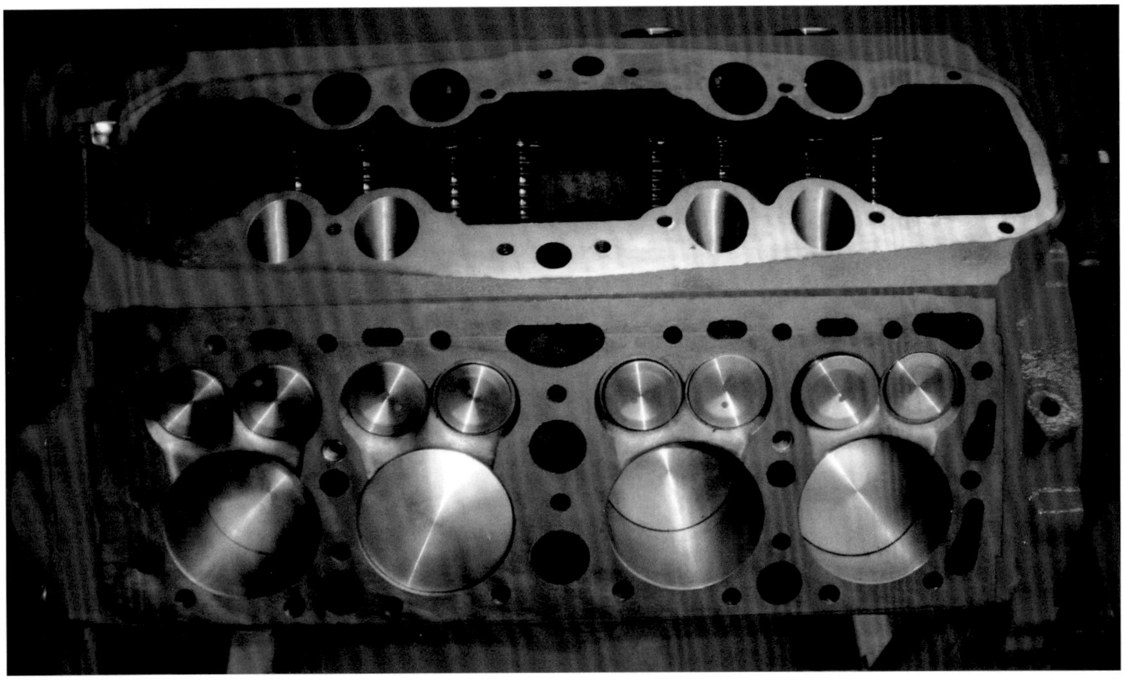

Figure 3-1. Flathead "Short" Block. For the flathead, this obviously includes all the valve train components as well as the intake and exhaust ports. Maybe that's why they are so expensive!

SUMMARY

Although individual parts are important, the proper preparation and combination of these parts determines what makes an outstanding engine versus a mediocre engine or even a catastrophic failure. The following paragraphs will provide some insight into the choice and preparation of the major flathead engine block components. The end of this chapter addresses porting and relieving the block. The following chapters will discuss heads and induction systems. Various combinations of these parts, i.e. actual engines, along with test results are covered in the later chapters.

BLOCKS

Figure 3-2. *8BA* (1949-1953) Block with a Heavy-duty Center Main Cap. Installation requires align honing of the main bearing retainer bores. This is a Doug King cap. The center main carries twice the load of the front and rear mains, so money is well spent here and no oil pan modifications are required.

A "good" block is one that requires no crack repair and that can be reconditioned with a minimum of effort. With the oldest flathead V-8 blocks over 75 years old, it is getting harder and harder to find a good one. I prefer the newest blocks for this reason and also because there were improvements along the way. The 1949 – 1953 blocks are therefore the blocks of choice for a high performance engine, although any can be used. Table 2-1 recaps the specifications of all the blocks since 1938 for reference. The later or "***8BA***" type blocks definitely had the best cooling design of all the flatheads. See Appendix B. The ***8BA***'s also provide the best fit for a Roots type blower, since the water outlets on the head are up front, providing plenty of blower clearance.

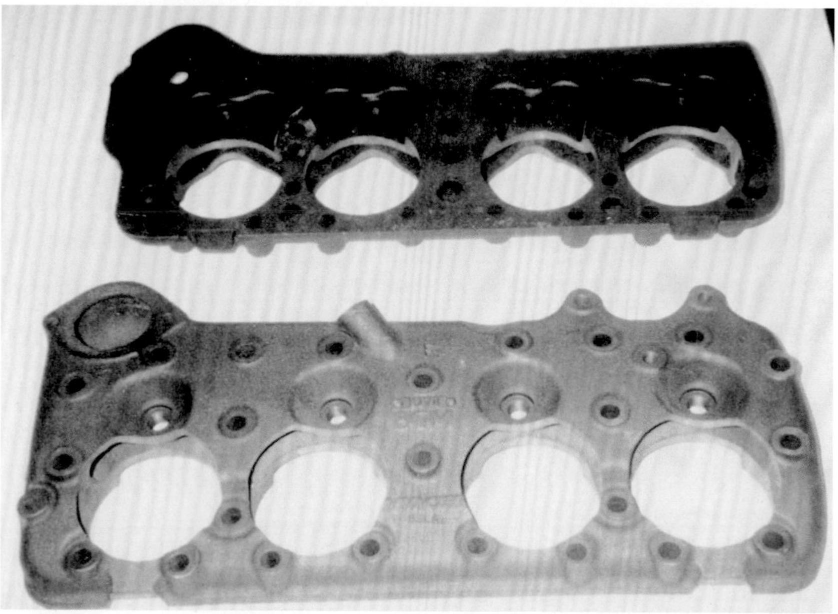

Figure 3-3. "Deck Plates" Made From Stock Heads. These are installed and torqued just as the actual heads during block machining. Main caps should also be installed during block machining.

Proper preparation should include degreasing, an acid clean to remove rust, followed by an acid neutralization dip. There are now several rust removal products that are still effective but safer than the Muriatic acid treatment outlined in ***Blown Flathead***. After thorough cleaning, the block should be thoroughly Magnafluxed and water pressure checked to detect cracks.

Figure 3-4. Main Cap Reinforcements. On the left is a heavy-duty steel cap. On the right is a stock cap with a very heavy duty "strap." Most straps are ½" flat steel.

Blocks with cracks in the valve pocket area or center main cap web area are not recommended for high performance applications. These cracks may be repaired by expert welding or pinning, but the repairs are often not permanent. **Cracks between water passages on the top of the block and head retaining bolt holes are common, but are usually not a concern and no repair is needed.**

Care should be taken to remove minimum material while "decking" the block since the decks are thin to begin with.

All threads should be "chased" or tapped. Marginal threads may be repaired with Helicoils or equivalent. If Helicoil-type inserts are used, thread sealer should be used on the back of the Helicoil threads to prevent corrosion and "weeping" of the head bolts or studs in service. **Thread sealer should also be used for head bolt or stud installation** for the same reason.

Boring and honing of the cylinder bores should be performed per the ring supplier's instructions. **Properly torqued deck plates and main caps should be used if possible during the boring and honing operations** to preserve cylinder roundness. This enhances break-in and ring seal in service. I have seen .001-.002" out-of-round bores when deck plates are not used. These can be made or rented if your shop does not have them.

MAIN CAPS

Heavy-duty main caps, supports, or full "girdles" are often used to strengthen the flathead crankshaft support. **Mild street engines do not require any support**. I have used a strap or a heavy-duty center main cap on my blown engines with no failures. The strap does not require align honing and is therefore the easiest to use. A heavy-duty main cap will require align honing. Since the center cap is loaded twice as heavily as the front and rear caps, I only reinforce the center cap.

PISTONS

There are numerous choices for pistons, including stock Ford units. A mild street engine can use cast aluminum pistons without problems. These are available in both 3 and 4-ring models and with flat top or domed top configurations. The 4-ring pistons are designed to provide greater stability of the piston in the bore and better oil control. However, the friction losses with a 4-ring piston may be unacceptable in a performance engine. Piston stability and oil control are not a problem with modern piston and ring designs. **Flat top pistons should not be considered for any application without a matching flat combustion chamber head configuration,** which is not common. A flat top piston with a domed cylinder head is extremely prone to detonation and defeats the Ricardo head design used in all high performance flatheads. See Appendix D.

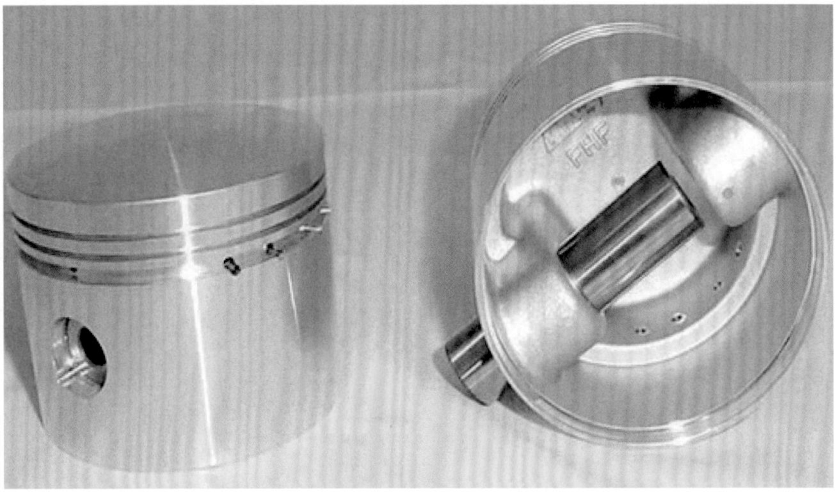

Figure 3-5. Ross Forged 3-ring Flathead Pistons. These are recommended for racing or hot street combinations. Mild street and low boost supercharged engines can use cast pistons. Piston-bore clearance is approximately double for forged vs. cast aluminum (e.g. 0.006" vs. 0.003").

The highest strength aluminum pistons are forged rather than cast. Forged pistons are recommended for all competition engines and for hot street engines where the piston acceleration, speed and combustion pressure are more extreme than those encountered in normal street driving. The forged pistons generally have larger bore clearances than cast pistons because of the higher expansion rate of the forged material, and thus can be somewhat noisier than their cast counterparts when the engine is cold. There are several suppliers for forged pistons including Arias and Ross.

In between these forged and cast pistons are the hypereutectic units. These are stronger than cast units but do not require the larger clearances of the forged units. Availability of hypereutectic pistons is lower than the cast units because of their higher tooling costs. I don't know of any current domestic source of hypereutectic pistons for the flathead.

RINGS

There are numerous choices for rings both in materials and design. The traditional flathead ring "pack" includes a 3/32" wide cast iron top ring, a 3/32" 2-piece second ring, a 3/16" 3-piece oil ring assembly, and a second 3/16", 1-piece oil ring (for a total of four rings). The recommended ring pack for a high performance flathead includes a 1.5 mm ductile iron top ring with a moly-plasma coating, a 1.5 mm second ring and a 3 mm, 3-piece oil ring assembly.

CONNECTING RODS

Even though the stock Ford rods look rather spindly, they are a very rugged forged unit and will do the job in any street application. They should be properly reconditioned and thoroughly checked for cracks before use. "Race prepping" the rods by carefully sanding the sides of the beams to remove the forging flash (reduces the stress concentration at the flash or ridge) and shot peening (to induce a compressive residual stress) increases the safety margin. New aircraft quality, grade 8 or better nuts should be installed. Locking "pal" nuts are not required if a thread locking compound is used during installation. The stock rod bolts are forged in place and cannot be replaced. There are several choices for high strength after market rods available if desired.

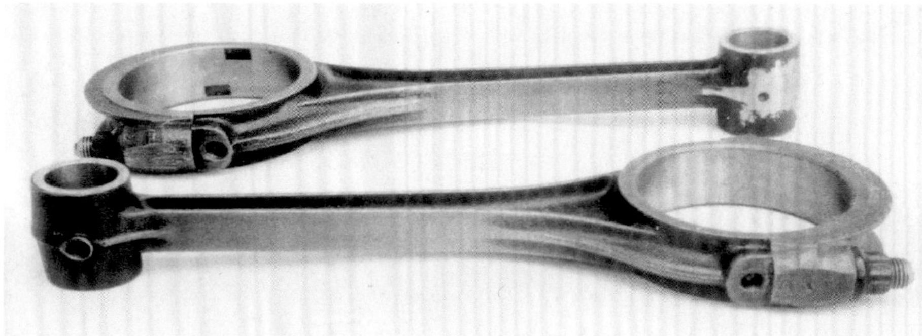

Figure 3-6. "Race Prepped" *8BA* Rods.

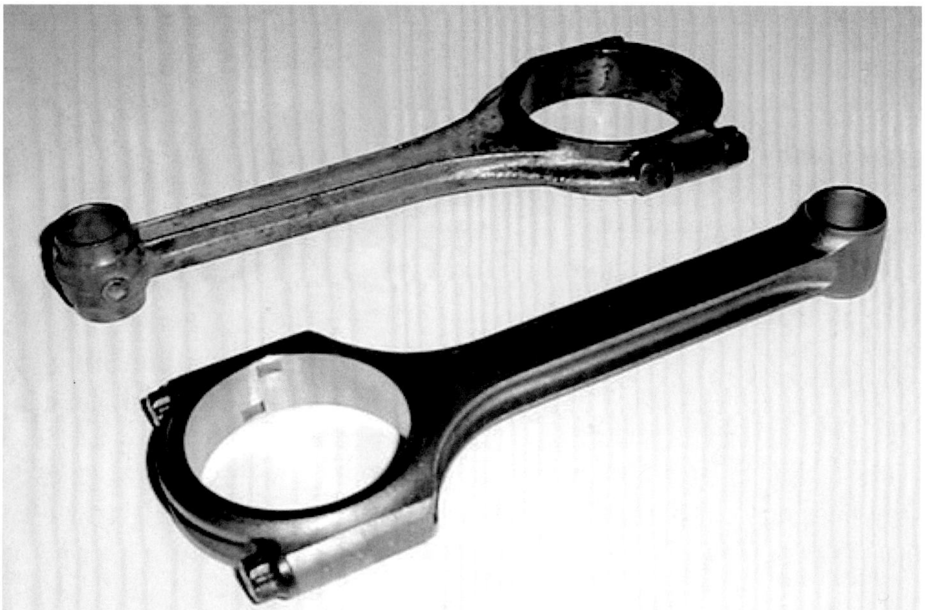

Figure 3-7. Stock *8BA* Rod and Scat H-beam Rod for 2.138" Rod Journals. The Ford rod is a very good piece and weighs in at 493 grams. The Scat rod shown is heavier at 547 grams.

Both I-beam (like stock) and H-beam units to mate with stock and undersize crankshaft rod journal sizes are available. These rods are larger and heavier than stock rods in most cases. In some applications, clearance grinding on the block is required to accommodate the larger size rods, particularly for strokes exceeding 4". Scat aftermarket H-beam rods are available to accommodate both the stock flathead journal size bearings (2.138") or the smaller 2" Buick-type bearings. To use the smaller bearings, the flathead crankshaft rod journals must be turned undersize to the 2" dimension. This results in less bearing drag (I estimate 3 hp less at 5000 rpm total all bearings), but also reduces crankshaft stiffness (about 20% less rigid). I recommend the stock rod journal size for all high performance street applications because of the additional crankshaft rigidity.

CRANKSHAFT

Figure 3-8. Flathead Crankshaft Related Parts Ready for Assembly. Stroker kits, including various combinations of parts, are now available from several sources.

Just as for rods and pistons, there are several options for the crankshaft. The stock Ford crankshaft from 1932–1953 has a stroke of 3-3/4 inches. The Mercury cranks from 1949–1953 had a 4-inch stroke and are generally preferred when available. The 1932-1934 cranks were forged steel while the later versions were steel castings. The later stock Ford and Mercury crankshaft rod journals were often ground undersize to 1.999" diameter, 1/16" off center, resulting in a 1/8" greater stroke. The earlier 21A rods with full floating bearings could then be used “as-is.”

The stock crankshafts are rugged units and will give good service in even the hottest street performance flatheads. Like the stock rods however, they should be thoroughly cleaned and magnafluxed for cracks prior to use. The clean-out plugs present on the later rod throws should be removed for thorough cleaning. New plugs should be a press fit and peened in place. A better option is to drill and tap for pipe plugs. A plug loss is instant disaster.

Figure 3-9. Stock ’49-’51 Mercury Oil Pump Pickup Assembly.

Scat produces cast 9000 series, nodular iron cranks, with strokes up to 4-3/8". At the top end are fully machined billet crankshafts from Moldex and others for all out racing applications.

OIL PUMP, PICKUP ASSEMBLY AND OIL PAN

The stock oil pump on a flathead is another very reliable item and rebuild kits are available. Brand new flathead units are available from Melling Engine Parts in two configurations. The stock equivalent is designated the ***M19*** and a unit with 20% greater flow capacity is designated the ***M15***. Both pumps are regulated with an internal by-pass valve to approximately 55 psi maximum. The high flow unit will maintain higher pressures up to the set point even when the engine is idling and hot, and when lower viscosity engine oils are used. The stock capacity pump will often only maintain pressures of 15 psi or less under hot idle conditions. The lower pressures are adequate at idle, and normally the pressure recovers at the rate of about 10 psi/1000 rpm, but these numbers are not very comforting to those of us used to modern engine parameters. The high volume ***M15*** unit is recommended for performance applications.

The oil pickup assembly must be compatible with the pan of course. The 1949 and up Ford car units and 1951 and later Mercs had the sump in the middle of the pan rather than at the rear like the rest of the flatheads. Where the ***M15*** pump is used, the oil pickup assembly must often be "massaged" slightly to get the proper clearance (about 3/8"-1/2") between the pickup screen and the bottom of the pan.

The pan should match the engine of course. The 1949 and later pans incorporate the rear crankshaft oil seal into the pan and thus are not interchangeable with the earlier units. The 1948 and earlier engines incorporate the rear crankshaft oil seal into the rear main bearing cap. My favorite pan, where it can be used, is the 1949-early 1951 Mercury pans, which have a windage tray stock. Custom pans are often fabricated from the stock pans for racing applications.

Figure 3-10. Stock 1949-1951 Mercury Pan Assembly. Note the built-in windage tray.

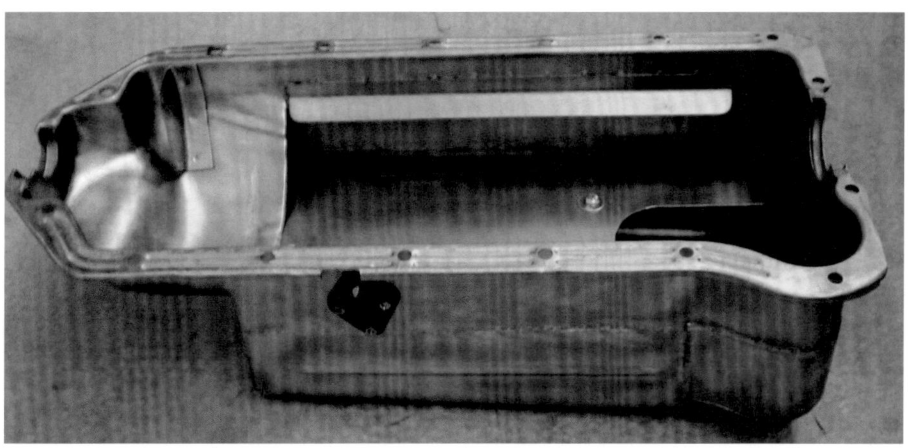

Figure 3-11. Custom Oil Pan Assembly for the *8BA* Type 1949 through 1953 Engines. The pan pictured was made from a stock oil pan assembly by Tom Porter of Whittier, CA.

CAM AND VALVE TRAIN

The flathead responds very well to a high performance cam. Indeed even the factory cams on the 1952-1953 Fords and Mercs were significantly upgraded from the earlier units with very positive results in power output. These later cams had a higher lift (.324" versus .296") and a longer duration (231 vs. 204 degrees) than the best earlier cams. The stock cams can be reground in a variety of profiles. Brand new cams on new cores are still widely available in a variety of old and new grinds. My personal favorites are available from Iskenderian and their dealers. The Iskenderian ***Max1*** is referred to as a "3/4 race" cam and provides excellent performance in a hot street engine whether normally aspirated or blown. The Iskenderian ***400Jr*** is a "full race" cam that is usable on the street, but produces significantly less torque at low rpm than the ***Max1,*** and has a very lumpy idle. The ***Max1*** will generally outperform the ***400Jr*** up to 4000–4500 rpm. There are numerous other grinds available from Isky and others. Even roller cams are available, but are usually not worth the money or effort for a street engine.

Stock lifters are very good items and the lightest available, but are non-adjustable. Valve clearance is set by sinking the valve or grinding the tip of the valve stem. This is quite tedious. Adjustable lifters, both solid and hollow, are also available. See Figure 3-12. The solid adjustable lifters are fairly heavy, but are good to 5500 rpm or more with single Isky ***185G*** valve springs, which is more than adequate for the street. The ***185G*** valve springs are at least equal to the Lincoln Zephyr units favored in the past. For valves, the stock 1.5" intake and exhaust units are usable. There is also a wide choice of new stainless steel units available. Small block Chevy exhaust valves will work in most applications and are available in head diameter sizes from 1.5" to 1.72" from most manufacturers. These valves are also available in various lengths, but are typically about .080" longer than the Ford units, which works out well with high lift flathead performance cams that have cam base circles .050-.100" smaller than stock. Custom valves are also readily available. I prefer 1.6" Manley

Figure 3-12. Flathead Tappets or Lifters. From left to right we have: 1) Stock, hollow, non-adjustable; 2) Unknown, non-adjustable; 3) "Old Johnson," hollow, adjustable; 4) "New Johnson," hollow, adjustable; and 5) "Solid," adjustable.

StreetFlo valves for both intake and exhaust. The 1.6" valves can be used with the stock hardened valve inserts or seats by simply grinding the seats to the appropriate diameter. The ***StreetFlo*** valves have two grooves on the stem, one for the keepers and one for the valve seals, for overhead valve applications. Either groove may be used for the keepers, thus easing adjustment of valve spring height and often eliminating the need for shims. I have found machined keepers to be necessary to achieve proper seating when using the seal grooves. The stamped keepers have a greater radius on the inner ridge, which will not fit properly in the valve seal groove on the valve, although they work fine in the actual keeper grooves.

MODIFYING THE BLOCK FOR BETTER BREATHING - PORTING AND RELIEVING

Since the intake and exhaust ports are in the flathead block, any porting and "relieving" must be accomplished before short block assembly. Any high performance flathead will benefit from a good porting job. Increases in intake airflow directly translate into increased power. The flathead is a very restricted flow design, but

Figure 3-13. Flathead Valves. From left to right: 1) Stock Ford Intake, 2) Stock Chevy-type Exhaust, 3) Manley *StreetFlo* Chevy-type Exhaust, and 4) Manley *ProFlo.* For 3), note the two grooves on the stem and that the backside of the valve is "undercut." These can be used for flathead intake and exhaust. The Manley *ProFlo* is similar to the *StreetFlo,* but has no seal groove on the stem and is made from higher temp material. These can also be used for flathead intake and exhaust. Note the Ford valve is about .08" shorter than the Chevy-type replacement valves.

responds very well to moderate porting. The intake and exhaust airflow can be increased 35% or more with a corresponding increase in power. Figure 3-17 and 3-18 illustrate the effects of porting. More flow data is provided in Chapter 4.

Refer to Figure 3-14. Opening up the heads over and around valves ("A" in the figure) is an easy and effective flow modification and will be covered in more detail in Chapter 4. Minor cleanup is satisfactory for the majority of the port diameter. Mild widening and cleanup in the "turn" area and rounding of the short side radius (upper side "B" in the figure) are recommended. High flow valves with undercut heads and stems ("D") are available from several sources. Blending of the bottom valve entry angle and the bowl area (below the valve) in addition to a cleanup or mild lowering of the hump near the valve guide ("E") are effective.

The merits of relieving (removing material at "C") a flathead have been vigorously debated in recent years. "In the day," all serious flatheads were relieved and with good reason. My flow testing of both relieved and unrelieved blocks has clearly demonstrated that relieving substantially improves flow. See Figures 3-17 and 3-18. High compression heads can show 14-18% airflow improvements with a moderate, .125" deep, constant-depth relief. Even low compression cylinder heads, with very large combustion chambers and large transfer areas (between the valves and cylinder bore), show 7-8% airflow improvements with relieving. **In the past the relief was**

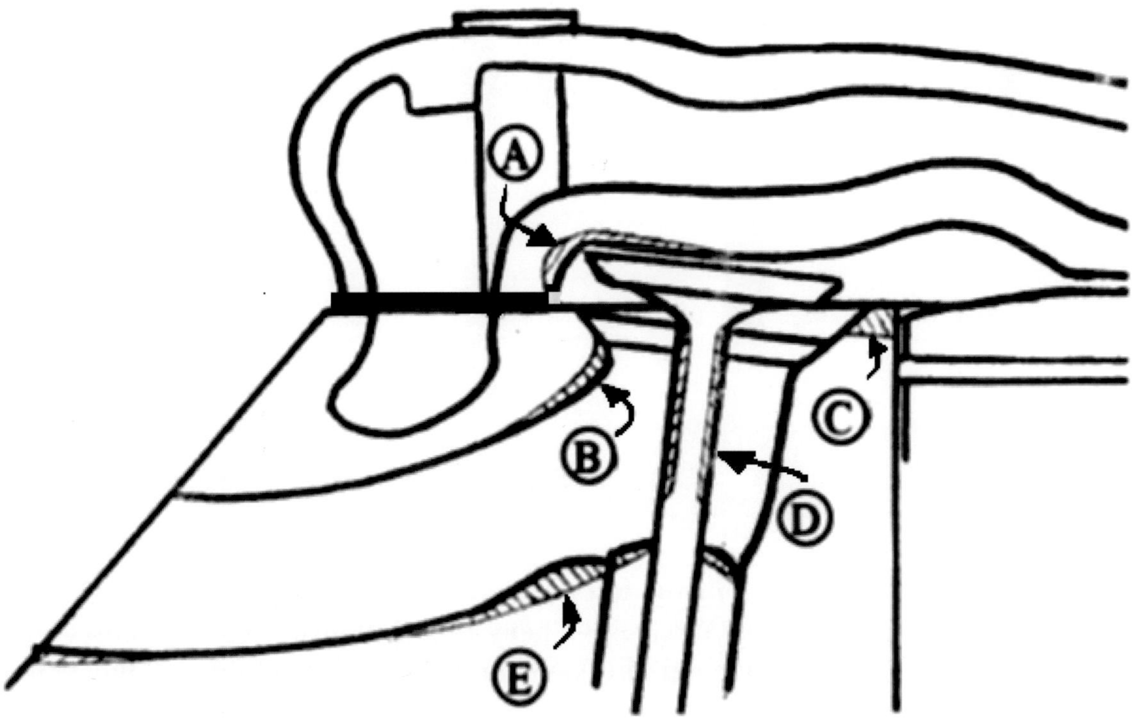

Figure 3-14. A Cross-section of the Flathead Intake Port, Block, Valve, and Head Configuration Showing Recommended Areas for Porting and Relieving.

often as deep as .187" at the cylinder bore. However, a .100"-.125" deep, constant-depth relief to the head gasket boundaries is nearly as effective and provides greater margin for grinding error and top ring clearance. See Figure 3-15. The constant depth relief flows substantially more than the ramp type (where the relief does not go into the cylinder bore). The only bad news is that relieving removes material from the combustion chamber and lowers the compression ratio. The good news is that not much material (about 3 cc) will be removed, which lowers the compression ratio only a small amount, typically 0.2 to 0.3 of a point. See Table 4-1 in Chapter 4 for determining flathead compression ratio. See also Appendix B of ***Blown Flathead***. Lowering the compression ratio 0.3 of a point only affects power about 1%, which is more than made up by the greatly increased flow that relieving offers.

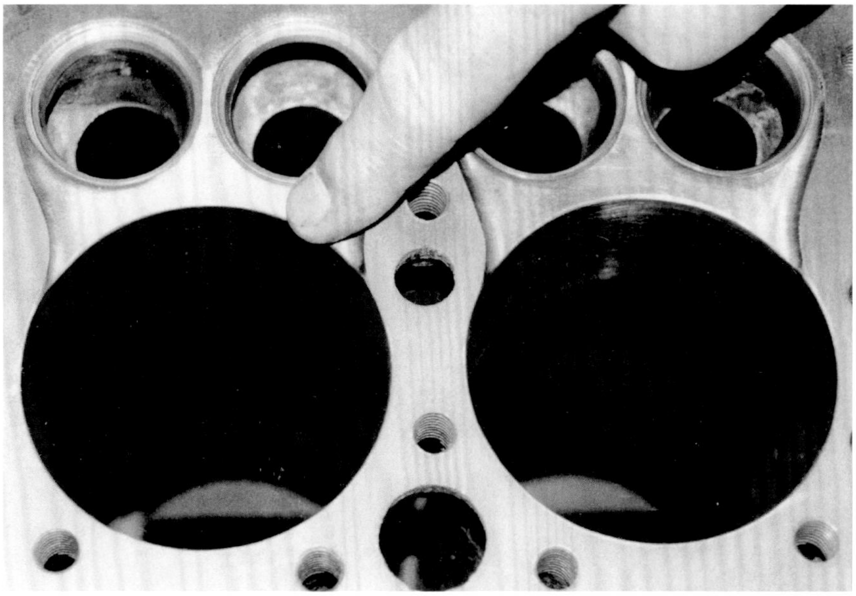

Figure 3-15. A Traditional Constant Depth Relief. Note the boundaries have a radius and although not visible in the picture, the intersection of the relieved area with the bore is chamfered or radiused where shown. *Jeff Andreen photo*

Some factory flathead blocks, like the "***59L***" Fords and the French blocks, have a relatively shallow and narrow factory relief. These reliefs do not substantially increase flow. They were reportedly added to reduce the cracking characteristics of hard working (usually overheated) truck and industrial flatheads.

A router is one method of relieving the block. See Figure 3-16. Joe Mondello told me he relieved 50 blocks a week in the 1950's using a Craftsman router! The routers only lasted about a month because of the metal cutting loads on the bearings, but hey, they were lifetime guaranteed! The performance advantage is attractive but relieving does require some skill, so some engine builders shy away. Give the job to someone who knows how or practice on a junk block first!

A good three-angle valve job is also a must as part of the port configuration. All flatheads from 1934 and up have hardened exhaust valve seats installed by the factory, and can be used without replacement for valves up to 1.6" diameter. Many of the second and third generation flatheads also have hardened intake seats. These are usually easy to spot visually or with your fingernail on the inside diameter area of the valve seats. The valve "pocket" or "bowl" (area under the valve seat) can be rough blended with the new larger valve seat with a "bowl hog" which has a 75 degree included angle. This essentially creates a four-angle valve job (30, 45, 60 & 75 degrees). Some builders prefer a 30 degree seat angle on the flathead intake valve seats rather than a 45 degree, but this is not recommended. The 30 degree seat does flow more at very low lifts, but most of the flow does not take place at low lift. The 30 degree configuration also does not provide as positive a valve seal, and does not center the valve as well as the 45 degree valve seat.

EFFECT OF BLOCK MODIFICATIONS ON FLOW CHARACTERISTICS

The following plots illustrate the very large gains in airflow and thus horsepower to be had by porting and otherwise modifying the flathead. More flow test results can be found in Chapter 4.

Figure 3-16. Here a router with a carbide cutter is being used to "relieve" the block. *Jeff Andreen photo*

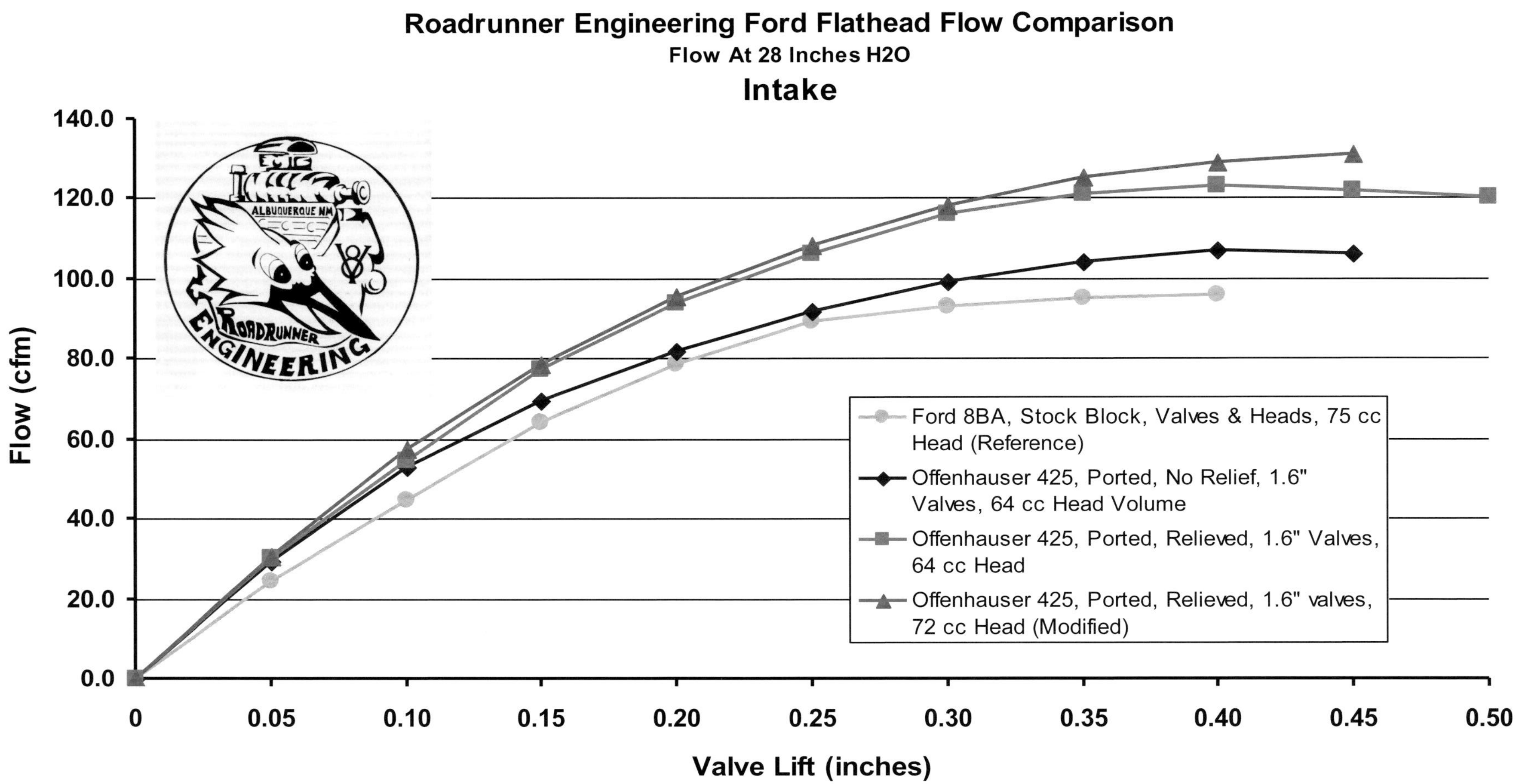

Figure 3-17. This figure illustrates that port modifications (including relieving) can increase airflow 35% or more over the stock configuration even with a typical high compression aluminum head. Two thirds of this increase is due to relieving the block! The test block utilized a traditional relief (see Figure 3-15) unless otherwise noted.

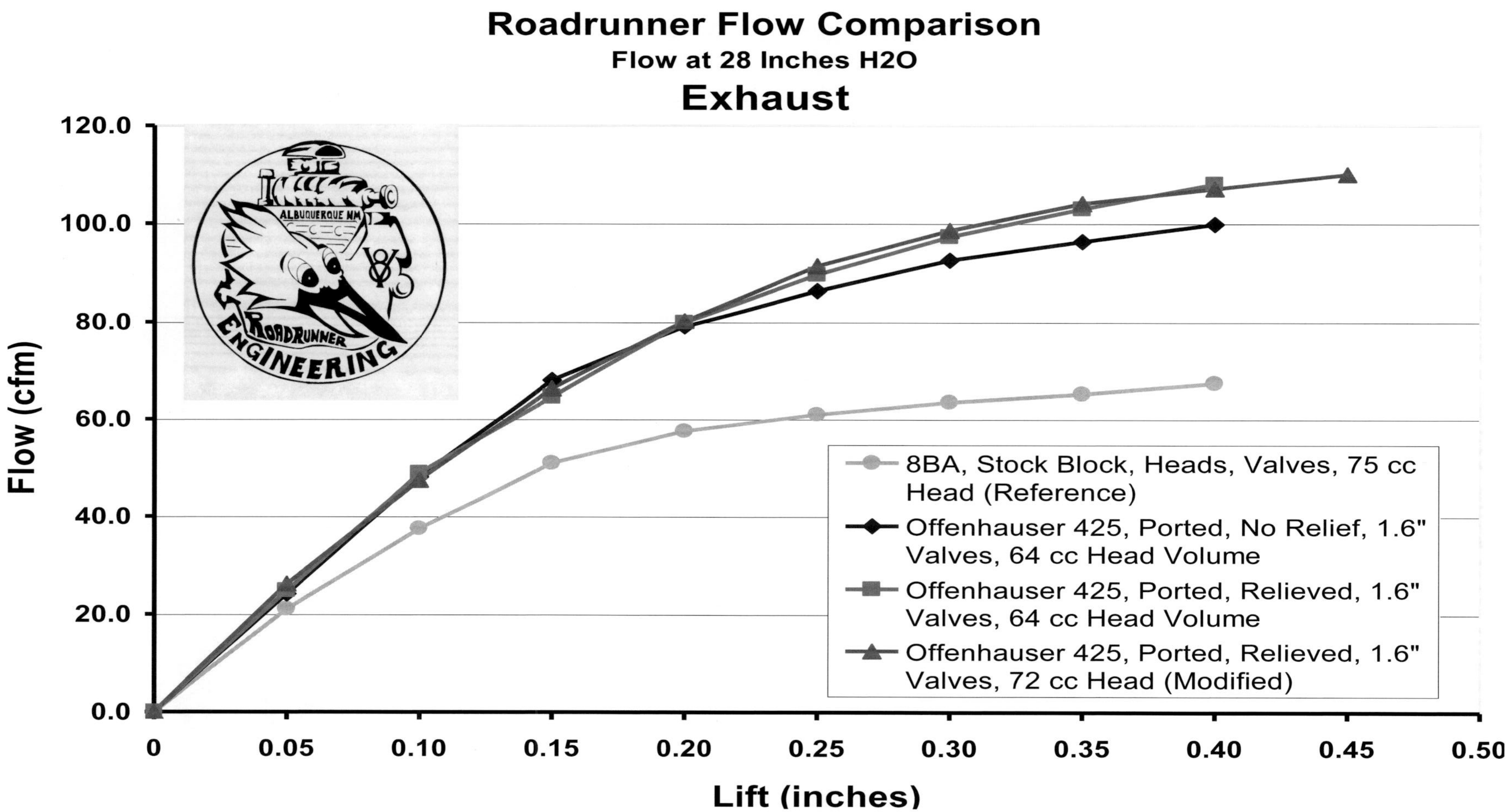

Figure 3-18. This figure illustrates that port modifications can increase exhaust airflow even more dramatically than intake airflow when compared to stock. Relieving the block is a smaller part of the increase when compared to that demonstrated with the intake, but it is still important.

-4-

Head Selection and Preparation

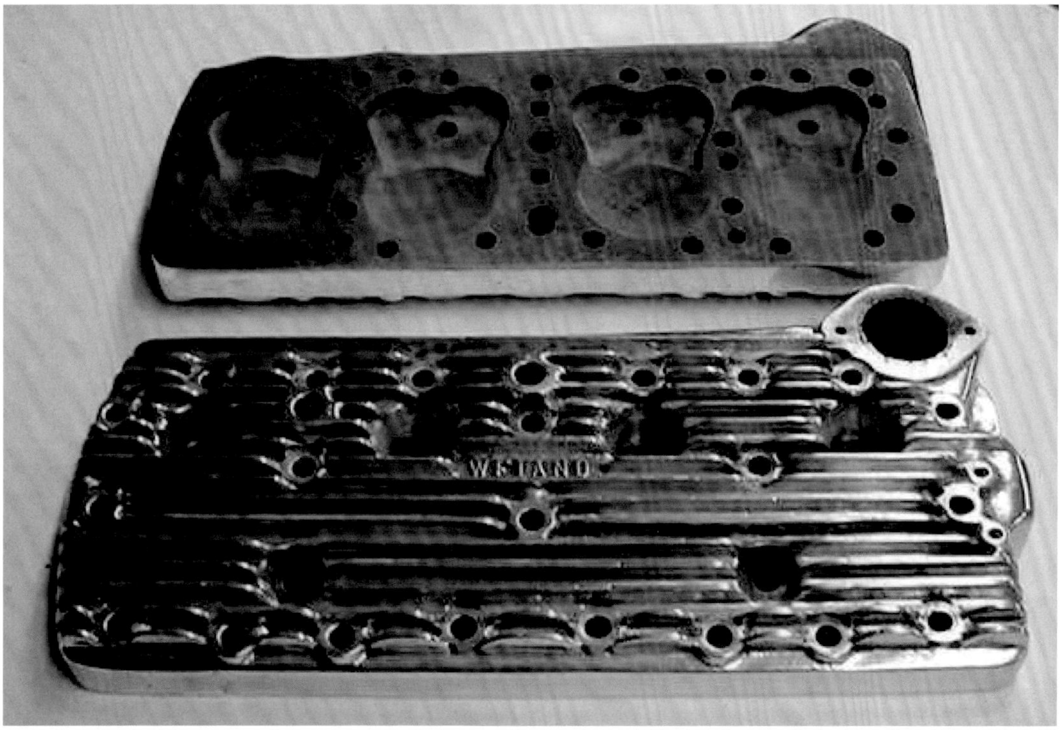

Figure 4-1. A Pair of Weiand Heads for the Late Flatheads. This was an excellent head, but is no longer being produced.

SUMMARY

In this chapter we will discuss the choice of heads for both naturally aspirated and supercharged engines. This is a three-step process where we determine the compression ratio and flow characteristics of the heads, and then evaluate their attributes using engine analysis software. In ***Blown Flathead***, a comparison of the pros and cons related to iron and aluminum heads for supercharged engines was presented. The iron heads are still an excellent choice for supercharged engines and give the best bang for the dollar, but performance flathead builders overwhelmingly prefer the aluminum heads for their traditional good looks. The aluminum heads also offer other advantages such as lighter weight, more design choices, and a greater resistance to detonation. Therefore, we will restrict most of our discussion in this book to the choice of aluminum heads.

THE TRADEOFF BETWEEN COMPRESSION RATIO AND BREATHING

The tradeoff between flathead compression ratio and breathing is discussed in detail in ***Blown Flathead***, and can be summarized as follows. Because of the flathead's unique valve placement next to, rather than over the piston (as shown in Figure 3-14), altering the flathead combustion chamber to increase compression ratio tends to reduce intake and exhaust flow, thus having a mixed effect on potential power output. Referring again to Figure 3-14, we note that if we "lower the roof" over the valves and the piston, we reduce the "transfer area" between the valves and the bore obviously restricting flow. Thus, we are faced with a tradeoff between the performance gains to be made with increased compression ratio vs. increased breathing capacity. To determine the optimum combination we have to determine the compression ratio and airflow characteristics of the head and block. Then, using modern analytical methods we can determine the highest performing configurations. The recommended procedure is detailed next.

THREE STEP METHOD FOR DETERMINING THE OPTIMUM HEAD DESIGN

The "best" combination of compression ratio and breathing characteristics for our flathead performance engine is determined as follows.

First, the combustion chamber volumes of the candidate heads are measured which allows the compression ratio to be estimated based on this volume and the other engine parameters using Table 4-1.

Second, the candidate head and block combinations are flow tested to produce data like that in Tables 4-2 and 4-3.

Third, the actual flow test data and compression ratio values are input into engine analysis software such as ***Engine Analyzer Pro*** to estimate engine performance. This

Figure 4-2. Cast Iron Mercury *EAC* Heads Being Modified for Additional Valve Clearance. This is often necessary when a performance cam is used with the iron heads. These heads are an excellent choice for supercharged engines because of their 76 cc combustion chambers.

Engine Displacement (cu.in.)		Head Combustion Chamber Volume (cc)							Engine Bore & Stroke
		88	82	76	70	64	58	52	
239	no relief	6.10	6.44	6.83	7.28	7.81	8.42	9.17	3-3/16 X 3-3/4
	relieved	5.95	6.27	6.63	7.05	7.53	8.10	8.78	swept volume 490 cc
249	no relief	6.31	6.67	7.07	7.54	8.08	8.73	9.50	3-1/4 X 3-3/4
	relieved	6.15	6.48	6.86	7.30	7.80	8.39	9.10	swept volume 510 cc
255	no relief	6.45	6.86	7.23	7.71	8.26	8.92	9.72	3-3/16 X 4
	relieved	6.28	6.62	7.01	7.46	7.97	8.58	9.30	swept volume 523 cc
265	no relief	6.67	7.04	7.48	7.97	8.56	9.24	10.07	3-1/4 X 4
	relieved	6.49	6.85	7.25	7.72	8.25	8.88	9.63	swept volume 544 cc
276	no relief	6.89	7.28	7.73	8.24	8.85	9.56	10.42	3-5/16 X 4
	relieved	6.71	7.08	7.49	7.98	8.53	9.19	9.97	swept volume 565 cc
284/286	no relief	7.08	7.49	7.95	8.49	9.11	9.85	10.73	3-3/8 X 4 or 3-5/16 X 4-1/8
	relieved	6.90	7.28	7.71	8.21	8.79	9.46	10.27	swept volume 582-586 cc
293/296	no relief	7.28	7.70	8.18	8.73	9.38	10.14	11.05	3-5/16 X 4-1/4 or 3-3/8 X 4-1/8
	relieved	7.09	7.48	7.93	8.44	9.04	9.74	10.57	swept volume 600-605 cc
304	no relief	7.49	7.92	8.42	8.99	9.65	10.44	11.38	3-3/8 X 4-1/4
	relieved	7.29	7.70	8.16	8.69	9.31	10.03	10.89	swept volume 623 cc

Notes:

1) Relieved assumes .100" - .125 constant depth cut in the block between the cylinder bore and the valves to the head gasket boundaries (traditional relief).
2) Assumes compressed head gasket thickness of .067" .
3) See ***Blown Flathead***, Appendix B for more information.

Table 4-1. Roadrunner Engineering Flathead Ford Compression Ratio Chart

procedure allows us to determine the effect of flathead head and block design on power without actual installation on an actual engine.

DETERMINING FLATHEAD COMPRESSION RATIO

As with overhead valve engines, the flathead's heads define a major part of the combustion chamber and the compression ratio. Unlike an overhead valve engine, the flathead has the intake and exhaust ports in the block as noted previously. Common compression ratio calculators for overhead valve engines cannot be used for the flathead, since they do not properly account for the transfer area volume between the valves and the cylinder bore. Table 4-1 allows us to quickly determine the compression ratio for most flatheads once we know the engine displacement and have measured the combustion chamber volume in the head. The table lists estimated values of compression ratio for both relieved (see Chapter 3) and stock blocks. The head volumes are given in 6 cc increments and we can estimate for values in between.

As an example, assume we have a 3-5/16 inch bore by 4 inch stroke flathead with a conventional 0.1" inch deep relief in the block. We measured the combustion chamber volume in the head as 64 cc. We are using a head gasket with an installed thickness of 0.067" (common). Referring to Table 4-1, we find that the compression ratio is about 8.5:1. Small deviations in relief depth or gasket thickness will not alter the results appreciably.

Determining the compression ratio by actual measurement (very difficult for a flathead) or for non-standard gaskets, pop-up pistons, etc. requires going back to the fundamental equations and procedures outlined in Appendix B of ***Blown Flathead.***

DETERMINING FLOW CHARACTERISTICS

The flow characteristics of the flathead are measured using a setup such as that shown in Figure 10-12. Obviously the test unit requires both the head **and** the block assembly. The Roadrunner flow test block incorporates both stock and modified test configurations. The "standard" test cylinder is "street" ported and relieved, and incorporates 1.6" intake and exhaust valves. The heads to be evaluated are then installed and the airflow measured vs. valve lift. Today, I primarily use a flow-measuring device called the ***Flow Quik*** by Audie Technology, which is easy to use, and portable. The results correlate very well with those obtained previously with SuperFlow ***300*** or ***600*** flow benches. Using the Roadrunner test block, I have flow tested over one hundred head and block combinations, and the results could easily fill a book by themselves. A representative sample of the Roadrunner flow tests of a stock Ford head and several aftermarket performance heads is shown in Tables 4-2 and 4-3. Some of this data is shown in graphical format in Figures 3-17 and 3-18 in the last chapter. See also Appendix A in ***Blown Flathead*** for more test results.

OPTIMUM COMPRESSION RATIO FOR A NORMALLY ASPIRATED FLATHEAD

As an example of applying the three-step head optimization method outlined previously, a stock Offenhauser 425 head was installed on the Roadrunner test block and flow tested with four different thickness head gaskets (to vary the transfer area and head volume). The calculated compression ratio and the flow test results were then input into ***Engine Analyzer Pro***. Adding head gaskets is **not** recommended for an actual engine because it promotes detonation by increasing the head to piston clearance. See Appendix D. Figure 4-3 summarizes the results of this study previously published in the December 2002 ***Street Rodder*** article, "Revival of the Fittest" (with corrections in the March 2003 issue). The analyses assumed a 276 cu.in. engine with an Iskenderian ***Max1*** cam ("3/4 race") and properly sized intake and exhaust systems, at sea level conditions. As shown in Figure 4-3, the highest compression engine dominated at lower engine speeds with higher torque, but then ran out of air at higher speeds. The lowest compression engine improved relative to the others as engine speed increased, but the improved flow never produced power equal to the others until the maximum engine speed simulated, which was beyond the horsepower peak.

These results indicate that **the optimum compression ratio for peak horsepower for a normally aspirated street flathead is about 8:1 to 9:1.** Going with the low side of this range has the advantage of re-

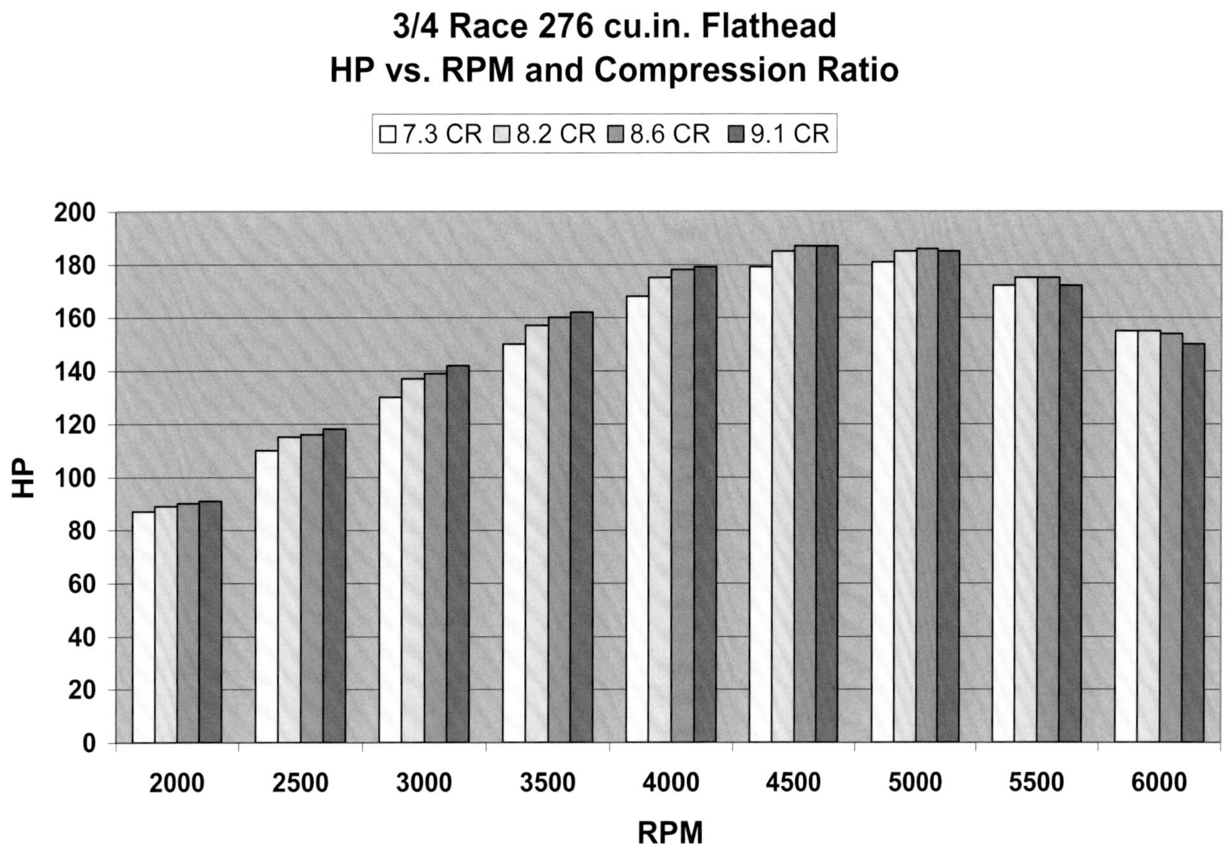

Figure 4-3. Estimated Horsepower vs. RPM for 276 cu.in. Flatheads. These engines are identical except for the head volume. More volume increased flow and lowered compression ratio and affected the peak power over the operating engine speed range as shown.

Manufacturer	Edelbrock	Edelbrock	Offy	Offy	Offy	Offy	Sharp	Weiand	Ford	Ford
Part Number	1115	1116	400	425 (no relief)	425	425 (72 cc)	59A	8BA	8BA	8BA (stock)
Test Date	4/28/03	9/18/08	4/29/03	1/15/00	1/11/00	8/22/06	4/28/03	5/17/06	6/12/03	1/5/97
Valve Lift (inch)	**Intake Flow in cubic feet/min (cfm) @ 28" H2O**									
0	0.0	0.0	0.0	0.0	0.0	0.0	0.0	0.0	0.0	0.0
0.05	29.0	29.5	28.6	29.4	30.2	30.6	29.1	31.2	30.7	24.3
0.10	53.6	54.4	54.6	52.9	54.6	57.4	53.6	56.2	54.8	44.7
0.15	72.6	74.3	74.1	69.2	77.2	78.6	72.9	75.3	73.1	63.9
0.20	88.5	90.9	90.1	81.6	93.6	95.2	90.6	90.9	90.2	78.5
0.25	103.0	106.0	103.0	91.8	106.0	108.0	104.0	106.0	106.0	89.0
0.30	113.0	118.0	111.0	99.2	116.0	118.0	115.0	119.0	117.0	93.1
0.35	119.0	127.0	114.0	104.0	121.0	125.0	122.0	126.0	122.0	94.8
0.40	120.0	132.0	115.0	107.0	123.0	129.0	125.0	133.0	123.0	95.8
0.45	119.0	134.0	112.0	106.0	122.0	131.0	126.0	134.0	123.0	
0.50		135.0			120.0			133.0		
Total Flow from 0.05-0.40 lift (cfm)	699	732	690	635	722	742	712	738	717	584
Maximum Lift (in.)	0.42	0.56	0.417	0.467	0.475	0.48	0.45	0.455	0.462	0.462
Head volume (cc)	60	74	54	64	64	72	63	69	76	76
CR @ 276 cu.in.	8.96	7.65	9.69	8.85	8.53	7.81	8.64	8.06	7.49	7.49
Weight (lbs)	13.2	13.6	16.6	15.2	15.2	15.2	17.4	15.2	24.5	24.5

Notes:

1) All tests used Roadrunner test block with "max street" port and 0.125" relief, 1.6" valves, except as noted.

2) All heads were factory stock except the Offy 425 (72 cc) which was opened up over the valves to 72 cc as described in Chapter 4.

3) 8BA (stock) had stock block with stock ports, valves, etc. in addition to the heads and is included for reference.

Table 4-2. Intake Flow Comparison for Several Flathead Ford Heads

Manufacturer	Edelbrock	Edelbrock	Offy	Offy	Offy	Offy	Sharp	Weiand	Ford	Ford
Part Number	1115	1116	400	425 (no relief)	425	425 (72 cc)	59A	8BA	8BA	8BA (stock)
Test Date	4/28/03	9/18/08	4/29/03	1/15/00	1/11/00	8/22/06	4/28/03	8/18/06	6/12/03	2/3/97
Valve Lift (inch)	**End Port Exhaust Flow in cubic feet/min (cfm) @ 28" H2O**									
0	0.0	0.0	0.0	0.0	0.0	0.0	0.0	0.0	0.0	0.0
0.05	24.8	24.9	23.8	24.2	24.8	26.2	23.0	27.3	25.6	21.0
0.10	49.5	47.1	47.4	48.0	48.8	47.4	47.3	49.4	50.2	37.5
0.15	70.2	65.4	63.6	67.9	64.6	66.2	66.9	69.4	67.3	50.8
0.20	85.9	78.3	75.6	79.0	79.9	80.2	80.1	83.4	78.5	57.5
0.25	95.4	90.7	87.0	86.3	89.6	91.4	90.8	93.8	88.9	61.0
0.30	101.0	98.9	92.6	92.5	97.3	98.6	98.8	102.0	96.7	63.5
0.35	105.0	106.0	97.3	96.3	103.0	104.0	105.0	108.0	103.0	65.2
0.40	105.0	112.0	98.2	99.8	108.0	107.0	110.0	112.0	105.0	67.3
0.45		116.0				110.0				
0.50		120.0								
Total Flow from 0.05-0.40 lift (cfm)	637	623	586	594	616	621	622	645	615	424
Maximum Lift (in.)	0.41	0.515	0.4	0.437	0.437	0.475	0.43	0.425	0.386	0.386
Head volume (cc)	60	74	54	64	64	72	63	69	76	76
CR @ 276 cu.in.	8.96	7.65	9.69	8.85	8.53	7.81	8.64	8.06	7.49	7.49
Weight (lbs)	13.2	13.6	16.6	15.2	15.2	15.2	17.4	15.2	24.5	24.5

Notes:

1) All tests used Roadrunner test block with "max street" port and 0.125" relief, 1.6" valves, except as noted.

2) All heads were factory stock except the Offy 425 (72 cc) which was opened up over the valves to 72 cc as described in Chapter 4.

3) 8BA (stock) had stock block with stock ports, valves, etc. in addition to the heads and is included for reference.

Table 4-3. Exhaust Flow Comparison for Several Flathead Ford Heads

Figure 4-4
Flathead *Engine Analyzer Pro* Simulation
3/4 Race 276 cu.in. Comparison
Relieved vs. Not Relieved

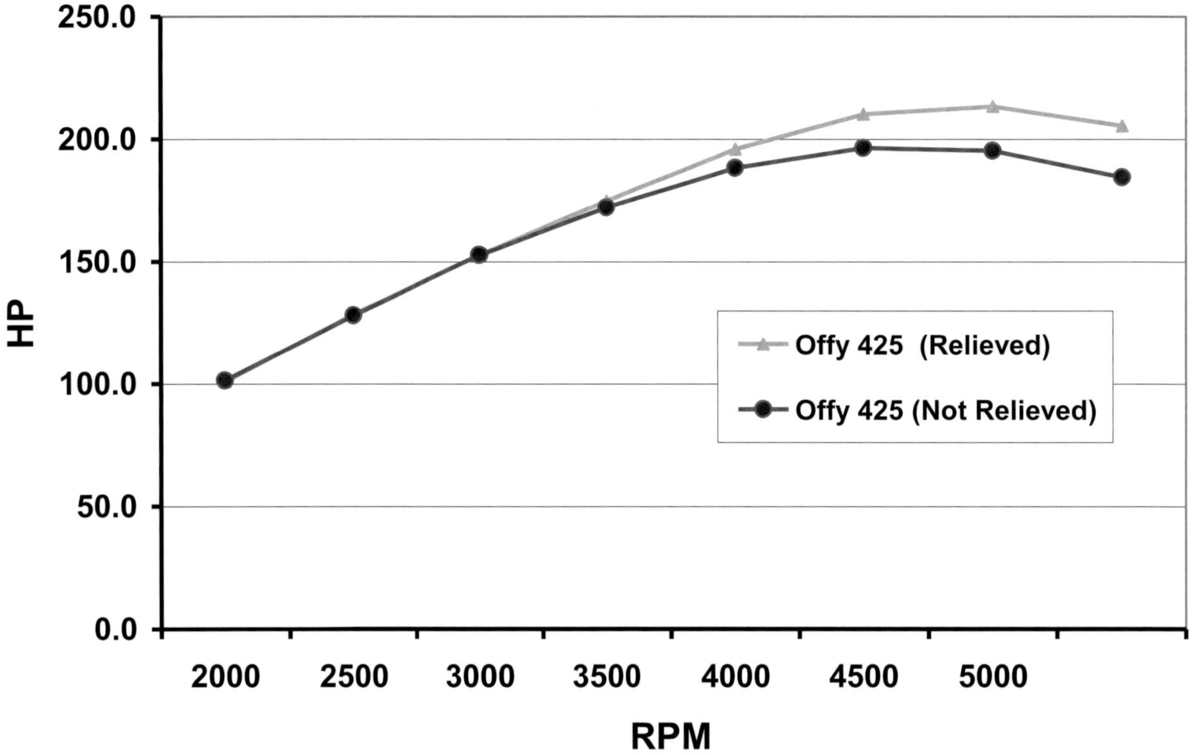

quiring lower octane fuel with minimal loss of low-end torque and mileage. Years of street driving experience validate these results.

Supercharged street flatheads work best with compression ratios somewhat lower. More on supercharged engines later.

A second study, which used the three step method, was used to evaluate several commercially available aluminum flathead heads provided by ***Street Rodder***. The results were documented in an article in the July 2004 issue, "Out with the Bad Air, In with the Good." The flow test results and the head combustion chamber volumes for the top performing Edelbrock, Offenhauser and Sharp heads in that study are included in Tables 4-2 and 4-3. Plugging in the flow test data and head volume measurements from these tables into ***Engine Analyzer Pro*** allowed us to estimate the performance of each head on a simulated engine. As shown in Tables 4-2 and 4-3, the Sharp ***59A*** and the Offy ***425*** heads had virtually identical measurements, and their simulated engine performance was likewise nearly identical. Thus only the Offy ***425*** data are plotted in Figure 4-4. The data from the Offy ***425*** tested on the ported but not relieved block, are also plotted for reference. This again illustrates the significant performance gains to be had with relieving. The top rated Sharp, Offenhauser and Edelbrock heads were very close in their estimated performance characteristics and all would have made a good choice for a **normally aspirated** street flathead.

The Sharp head was chosen for the actual ***Street Rodder*** flathead buildup of an early (pre-1949) engine and performed very well, as expected. Unfortunately, this head is not available for the later engines.

Many of the top performing heads, such as the Offenhausers and Edelbrocks, are available for all flatheads from 1938-1953. These heads have stood the test of time and have been available continuously for over 50 years. With the resurgence of interest in the flathead, other old names in aluminum heads in addition to the Sharp, such as Eddie Meyer and Cyclone, have been resurrected. There are also new designs available, such as the Baron, Flatattack, and Wilson & Woods heads. Other brands produced in the past include the Weiand, Evans, Edmunds, Fenton, Grancor, Kong, Champion, and many others. See photos of some of these heads below.

A SAMPLING OF FLATHEAD HEADS

Figure 4-5. Edelbrock *1125* head for the early (pre-1949) flatheads. This head was evaluated by Roadrunner in 2002 and the data shown in Table 4-2 and 4-3. The combustion chamber is shown in Figure 4-7.

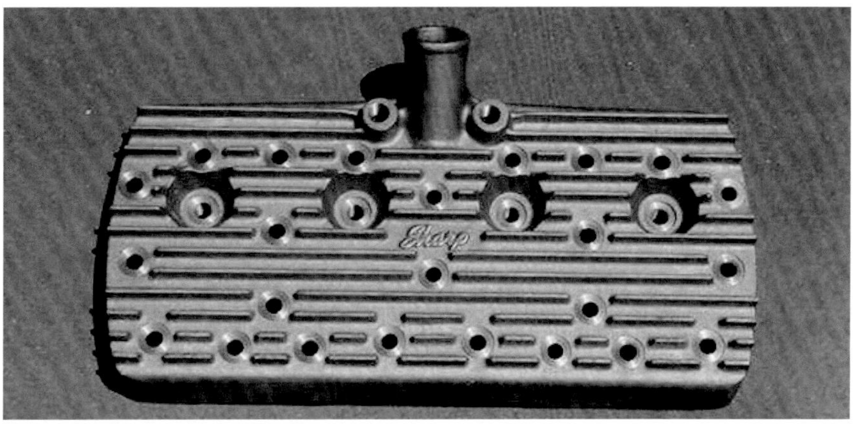

Figure 4-6. Sharp *59A* head for the early (pre-1949) flatheads. This head was evaluated by Roadrunner in 2002 and the data shown in Tables 4-2 and 4-3. The combustion chamber is shown in Figure 4-7.

Figure 4-7. Shown from left to right, are Edelbrock *1125*, Offenhauser *400* and Sharp *59A* heads showing the combustion chamber configurations. The flow characteristics for these heads are shown in Tables 4-2 and 4-3.

Figure 4-8. A Navarro head for the 1948 and earlier flatheads (*59A* type) is shown at the top of the photo. An Offenhauser *425* head for the 1949 and later flatheads (*8BA* type) is shown at the bottom. Both brands are available for both early and late engines. The combustion chambers are shown in Figure 4-9.

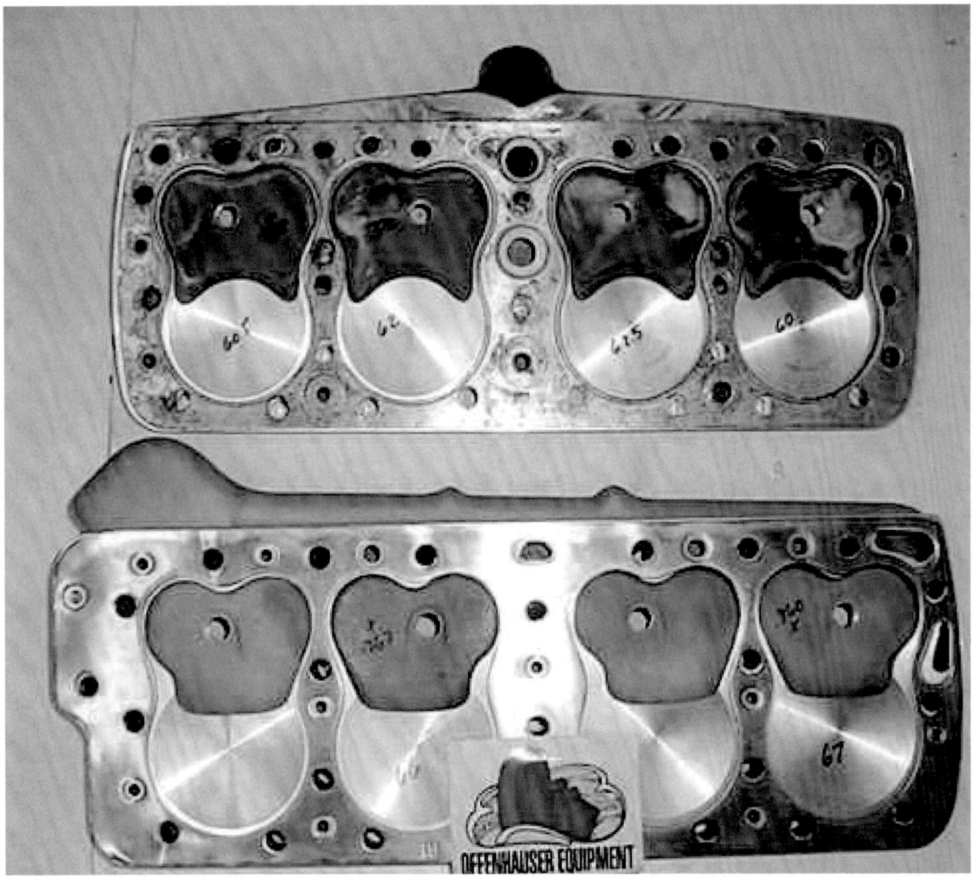

Figure 4-9. Here we see the combustion chamber side of the same Navarro (top) and Offenhauser (bottom) heads as shown in the previous figure.

Figure 4-10. The Baron Racing Head. These have a fully machined combustion chamber and can be ordered with a variety of combustion chamber configurations including versions to properly accommodate flattop and popup pistons.

Figure 4-11. A new head from Wilson & Woods does not incorporate fins as most other aluminum heads. The pictured head has been engraved with the customer's marking.

HEADS AND COMPRESSION RATIO FOR SUPERCHARGED ENGINES

In the December 2002 ***Street Rodder*** article, it was noted that a racing flathead might require a different head than the normally aspirated street engine featured in the article. A higher operating rpm range would favor better breathing, lower compression heads for peak power. The use of a supercharger, nitrous or racing fuels would likewise demand lower compression ratios for longevity in extreme racing conditions. Poor fuel economy would generally not be much of a factor. **Thus, it is important not to overextend the results of our normally aspirated street rod head selection process to other applications.**

Cylinder pressure and heat are what make power. The trick is to maximize these power producers over time such that we stay within the structural limits of the engine. Excessive pressure and heat can cause major structural problems such as splitting cylinders, collapsing pistons, bending rods, breaking crankshafts, etc. Likewise excessive pressure and heat can contribute to lesser problems such as blown head gaskets, high wear and shortened engine life. Ideally what we would like for peak power is a high, but still safe, cylinder pressure that stayed constant over the entire power stroke. This ideal can be somewhat approximated by controlling the rate of burn and the expansion process. Detonation is an uncontrolled burn (an explosion) that kills power, if not the engine itself, and therefore should always be avoided. Detonation results in very short duration, extremely high pressure pulses that can cause severe engine damage, sometimes within fractions of a second.

As it turns out, a controlled steady burn and lower compression ratio (equates to a lower expansion ratio also) approaches our ideal process for maximizing power without breaking parts. Most **normally aspirated** engines with compression ratios less than about 8.5:1 will **not** be operating in the detonating mode or at an excessive cylinder pressure when properly tuned with today's pump gas. Supercharged engines however, can easily exceed safe cylinder pressures when high boost levels are combined with high compression ratios, even if they are not detonating. The peak cylinder pressure is much more sensitive to compression ratio than boost pressure. **Using the three step process, and evaluating heads for <u>supercharged</u> engines, demonstrates that we can make more power safely with higher boost and lower compression ratios, rather than the other way around**. See examples in Table 4-4.

Lower compression ratios have the disadvantage of lower fuel efficiency and hotter engine and exhaust temperatures. Thus for a **supercharged** street engine where fuel economy, overheating and exhaust valve survival are important, we should keep the compression ratio in the range of about **7.2:1 to 8.2:1**. Supercharged race engines may sometimes approach compression ratios as low as 5.5:1 with boost pressures of 20 psi or more for very short duration (e.g. drag racing).

In summary, numerous studies and actual engine experience have convinced me that **supercharged flathead compression ratios of about 7.2:1 to 8.2:1 provide a reasonable compromise for good power and fuel economy** with today's 91-93 octane pump gas. Aluminum heads can operate on the high side of this range and somewhat higher since they are less prone to hot spots and detonation due to aluminum's higher conductivity compared with cast iron.

MODIFYING PRODUCTION HEADS FOR BETTER BREATHING AND LOWER COMPRESSION RATIO

If we refer to Table 4-1, we note that most commonly available aluminum heads have a higher than desired compression ratio for supercharged flathead engines in common displacements. In many cases, this can be remedied by removing material from the heads in the valve area. **Do not remove material from the piston dome area of the head or use multiple head gaskets since these measures actually increase the tendency toward detonation.** See Appendix D.

Compression Ratio	HP	Boost (psi)	RPM
7.5	274	6.2	5000
7.9	263	4.9	5000
8.9	246	4.4	5000

Table 4-4. Estimated power output vs. compression ratio for three otherwise identical high performance supercharged flatheads. Boost was the maximum that could be tolerated without detonation using 91-93 octane fuel.

Key point: more boost and lower compression ratio makes more horsepower safely. Moderation is advised!

Removing material from the valve area of the head can be accomplished effectively by using a 1.875" diameter end mill with a 0.125" corner radius. See the following figures. Using the head gaskets that we will use during assembly as a template, we can safely un-shroud the area in back of the valves to the gasket boundaries. We can also usually safely remove .060-.125" from the "roof" of the head over the valves to add about 4-8 cc to the head combustion chamber volume. The following photos show the modification of an Offenhauser ***425*** head, which has a relatively thick combustion chamber wall thickness. Not all heads are the same however, and care should be taken to maintain a minimum head thickness of .375" or more, since supercharging can easily double the peak combustion pressures.

Figure 4-12. Here an end mill is being used to remove material from the head over the valves. The face of the cutter should be parallel to the gasket surface of the head for maximum enhancement of breathing.

Figure 4-13. This photo shows the result of a .060" end mill “plunge” from first contact of the cutter. Note .125" corner radius.

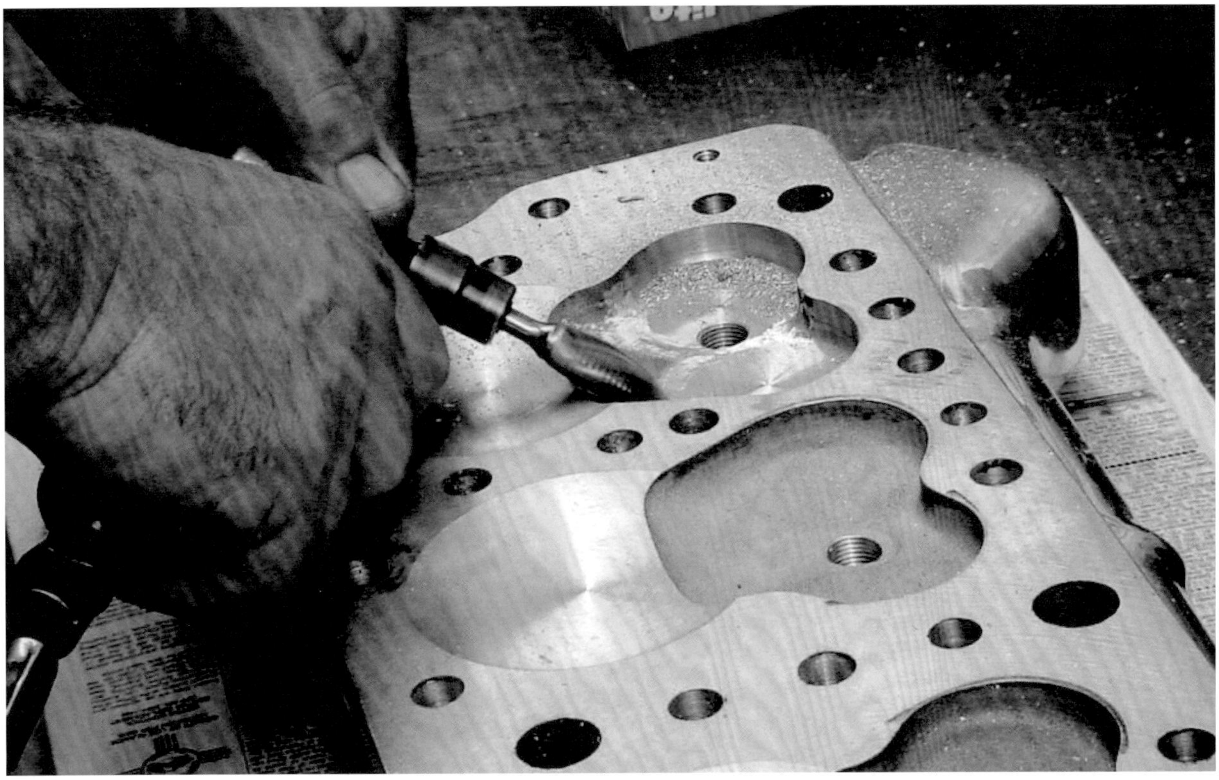

Figure 4-14. Here the end mill cuts are blended using a flame-shaped porting tool. An abrasive "flapper" wheel is used to finish the job.

Figure 4-15. The Head After Our Combustion Chamber Enlargement. The final head combustion chamber volume was approximately 72 cc.

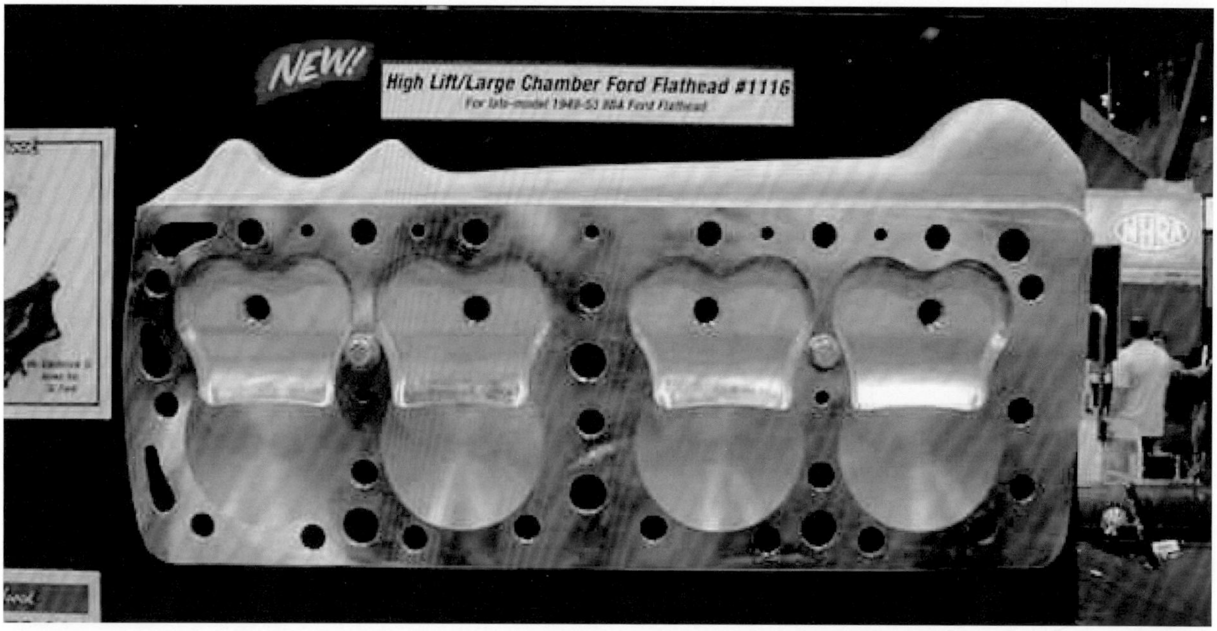

Figure 4-16. This is Edelbrock's big chamber (74 cc) heads for the *8BA* engines, P/N 1116. The 74 cc combustion chamber will yield a compression ratio that is a good match for many supercharged applications. Use Table 4-1 to determine your compression ratio. Flow data for this head is included in Tables 4-2 and 4-3.

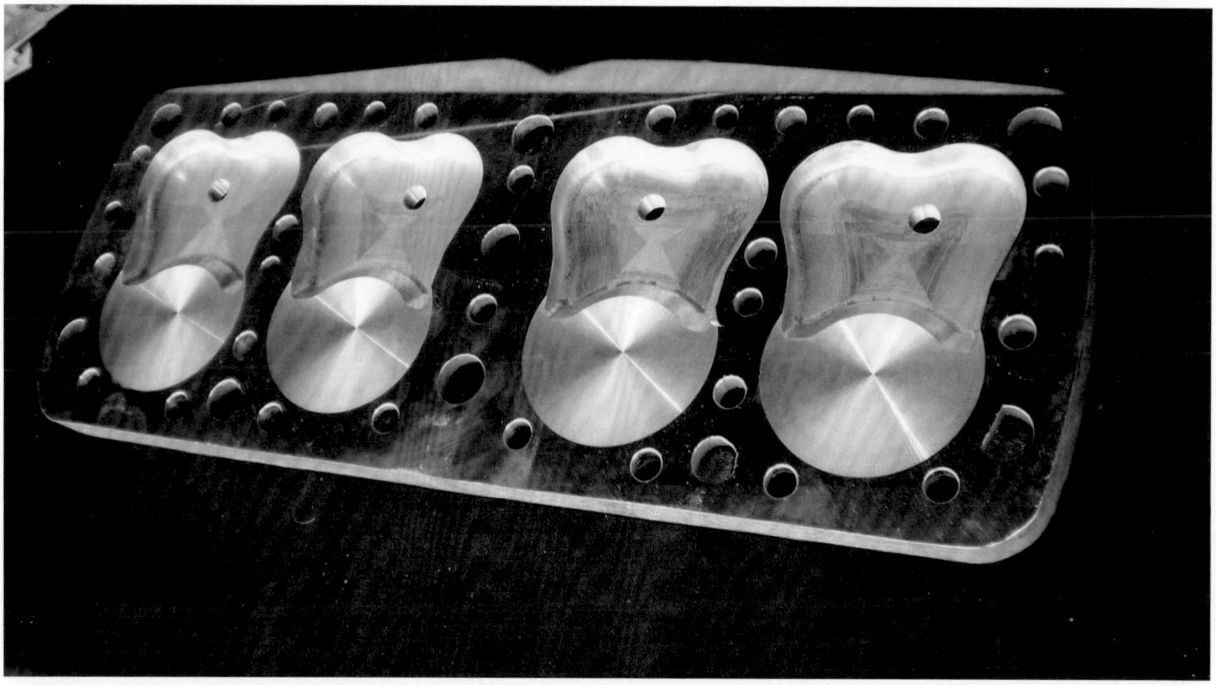

Figure 4-17. Navarro 75 cc Head For The *59A* Engines. This is a new head and has not been evaluated by Roadrunner Engineering.

-5-

Normally Aspirated Flathead Induction, Exhaust, and Ignition Systems

Figure 5-1. A Very Rare Eddie Miller Manifold on "Doc" Parson's Model T Street Rod from Tulsa, Oklahoma. The manifold mounts three Stromberg *E* carburetors (1-bbl) and one Stromberg *EE* carburetor (2-bbl)! It dates from 1935.

SUMMARY

In this chapter we will take a general look at various flathead intake, exhaust and ignition system options. The stock pieces could hardly be classed as performance items but were adequate for their intended use, and were somewhat improved over the flathead's lifetime by the factory. This chapter includes factory and Roadrunner test results of an essentially stock ***59A*** flathead tested on an engine dynamometer with several variations and combinations of bolt-on intake and exhaust systems. The best performing bolt-on intake and exhaust systems tested increased power by 20% on this engine.

At the end of this chapter, test results are presented for an internally modified ***8BA*** flathead with several intake systems and tuneups. The results give a preview of information presented in later chapters that demonstrates that flathead horsepower can often be doubled normally aspirated, and even tripled with supercharging.

STOCK CARBURETORS AND INTAKE MANIFOLDS

The stock flathead V-8 induction systems came in several configurations over the lifetime of the engine. The 1932 and 1933 versions had a 1-barrel Detroit Lubricator carburetor and an essentially open plenum aluminum intake manifold. In 1934, the intake system was redesigned to use a 2-barrel Stromberg Model ***40*** carburetor and the manifold itself was redesigned to be a 180-degree dual plane design. A dual-plane manifold is essentially two manifolds in one, with the ports arranged such that each barrel of the two carburetor barrels feeds half of the cylinders, and the ports are alternated such that no cylinder draws from the same barrel as the previous firing cylinder.

Figure 5-2. High rise Tattersfield 2x2 intake manifold and swept back tubular steel exhaust headers are shown on an early flathead. These look good but there are better performance options for both the intake and exhaust.

In 1935, the Stromberg Model ***48*** carburetor replaced the Model ***40***. The Model ***48*** was replaced in 1936 by the famous Stromberg Model ***97***, which was used through early 1938. While these carburetors are prone to leaking and flooding, they are easy to tune and they maintain an excellent air-fuel mixture over the operating range. The flooding issues can be managed with modern float valves and a good fuel pressure regulator that allows no more than 3.5 psi of fuel pressure.

Ford used various versions of the Chandler Groves or Holley Model ***94*** from 1938-1953. Dry airflow capacity of several of these carburetors is shown in Table 5-1. Wet flow capacity is approximately 8% less.

Carburetor	Type / Model	Airflow (cfm) @ 1.5 in. Hg	Comments
Stromberg	48	183	
Stromberg	81	114	
Stromberg	97	163	Average of 3
Holley	94	153	
Holley	1CM, 8CM	210	1949-51 Mercury

Table 5-1. Airflow Comparison of Selected Vintage Two-Barrel (2-bbl) Carburetors.

The Stromberg carburetors listed in Table 5-1 all have manual power valves and no vacuum ports. The Holleys all have vacuum operated power valves with various operating points available. While the vacuum operated power valves work well with single carburetor applications, they can complicate multi-carburetor tuning.

Mercury used four-bolt base carburetors of several models manufactured by Holley from 1949 through 1953. These carburetors had good airflow capacities, but are not commonly used for performance applications.

HIGH PERFORMANCE INTAKE SYSTEMS

There have been and still are literally hundreds of flathead intake systems available, both new and used. Many are still in current production from major manufacturers. The performance of several representative flathead normally aspirated and super-charged engines are presented later in this book for reference.

Normally aspirated, a 3x2 intake system such as shown in Figure 5-3 will give very good performance, drive-ability, and economy. A 390-570 cfm four-barrel carburetor with vacuum operated secondaries will also provide good results. See Table 5-2 at the end of this chapter. No matter what carburetion system is chosen (3x2, 1x4, etc.), there are definite performance differences in manifolds and carburetors. Unfortunately, there is no way to know which is best for a particular application without testing.

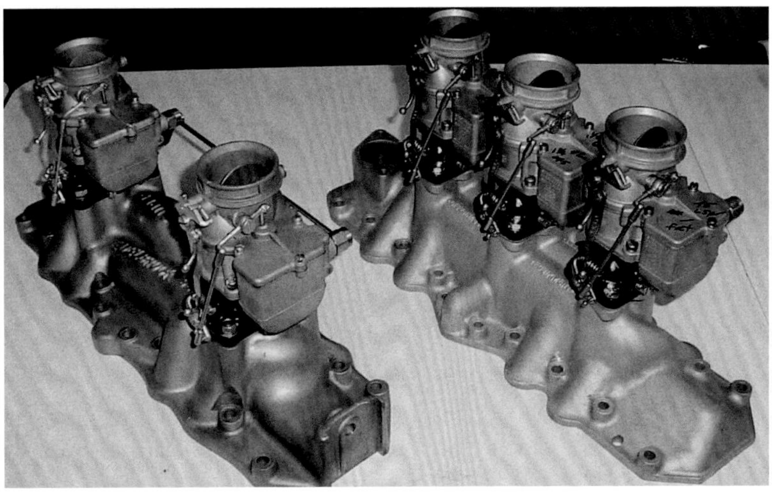

Figure 5-3. Two excellent flathead intake manifolds for the early (*59A* style) flatheads. On the left is an Edelbrock *Super Dual* and on the right is an Offenhauser *Triple*. Both mount Stromberg *97* carburetors. Both setups were evaluated and the results reported in later chapters. Note both manifolds shown require offset generator mounts.

FUEL INJECTION

Fuel injection options in the '50's for the flathead were pretty much limited to the Hillborn or similar mechanical systems. These systems were competition only systems that worked very well for their intended applications. In recent years, there have been several electronic fuel injection (EFI) systems adapted for use on street flatheads. These include units produced by Hillborn, Holley, Moon Equipment, Howell Electronics, Flatattack, Momar Injection, Flathead Engineering and many others. All of these systems have been used successfully on the flathead. That being said, none of these systems are usable without significant test time and tuning by a knowledgeable technician. When setup properly, the EFI systems can provide economy, drivability, and power equal to or better than any carburetion system. However, carburetors still provide the best bang for the buck, and are much more user friendly than currently available aftermarket EFI systems. Even though carburetors have not been used on new production vehicles for more than 20 years, they are still widely available and provide excellent service on the street and on the track. For detailed EFI user information the reader is encouraged to contact the manufacturers. See Appendix A.

Figure 5-4. Hillborn injected flathead dragster. This is a mechanical unit without any electronics available essentially unchanged since the '40's. Note fourth exhaust port exiting from the former manifold heat riser port. Hillborn currently produces an electronic version of their fuel injection for the street, as well as the mechanical competition unit shown.

STOCK EXHAUST SYSTEMS

The stock flathead exhaust manifolds have to be candidates for the all time awards for most restricted and worst appearing. Many, like the 1932 exhaust manifolds, were simple log-type cast pieces that bolted to the block with no significant transition or streamlining, and were connected to the tiny exhaust pipes. On the 1932, the tailpipe was only 1.5" in outside diameter to further restrict any exhaust that escaped through the manifolds! The Mercurys from 1949 on, probably had the best stock exhaust systems with larger manifolds, and large straight-through mufflers that even

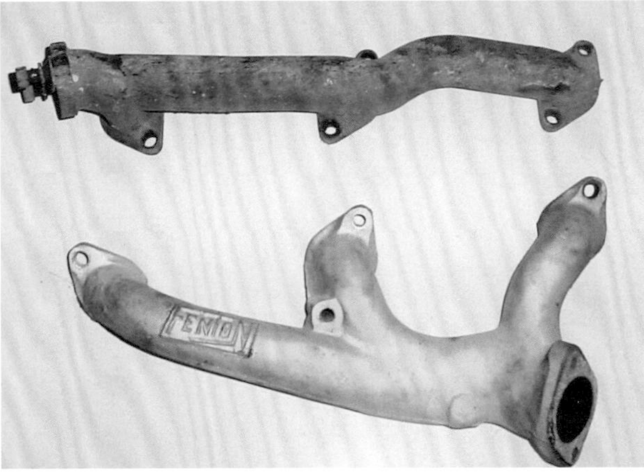

Figure 5-5. A stock flathead driver's side exhaust header is shown above the equivalent Fenton header. The stock outlet faces forward and attaches to a crossover pipe to the passenger side header (single exhaust). The Fenton is a good unit with a stock appearance and longevity.

sound pretty good. Some 1953 Fords even came with dual exhausts, and the driver's side manifold for these cars has been reproduced sporadically.

HIGH PERFORMANCE EXHAUST SYSTEMS

Luckily, there are a large number of high performance options available for all flathead powered vehicles. The flathead exhaust breathing is so restricted in the block exhaust ports however, that unless the block is ported, the new bolt-on pieces are of limited improvement. Nevertheless every little bit helps. I favor the center- dump, Belond style (Pertronics in Figure 5-6) headers because the center exhaust port carries twice the flow of the end ports. Henry Ford chose the center-dump exhaust headers for his flathead powered Indy cars in 1935. Not all applications can accommodate the center-dump design because of interference with the wishbones in the early cars. Other designs in both cast iron and tubular steel are available to fit any application.

STOCK IGNITION AND TUNEUP

The flathead ignition systems used through 1948 all had front mount, direct cam driven distributors (shown on the left and center of Figure 5-7). All were 6 volt with a ballast resistor to limit coil current after the engine was warm and running. The ballast resistor was bypassed during starting to allow maximum current and voltage to the coil during

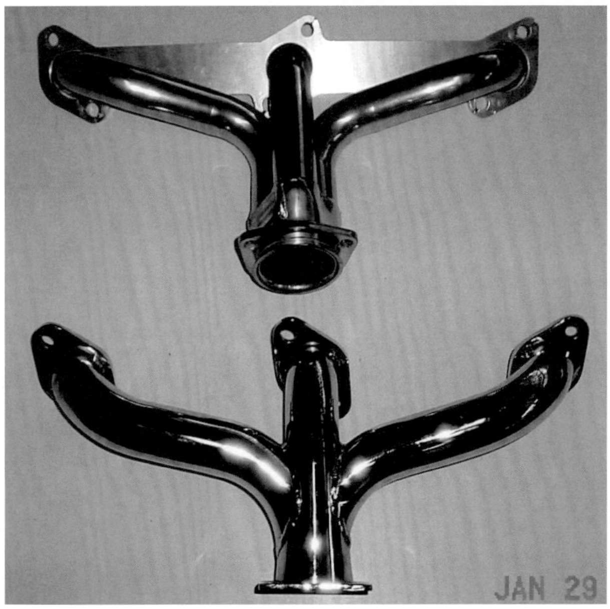

Figure 5-6. A Sanderson "center-dump" flathead exhaust header is shown on top, and a Pertronics center-dump is shown at the bottom. This style works very well on a flathead since it favors the center exhaust which flows twice the volume of the end exhausts on a flathead. The Pertronics header has given very good results in dyno testing but often requires significant "cleanup" of the interior passages and welds.

Figure 5-7. Evolution of the Flathead Distributors. On the left we have a "diving bell" type representative of those used from 1932-1941. In the center we have the "crab" type cap (1942) on a distributor representative of those used from 1942-1948. Both units are usable for performance applications. On the right we have a *Load-a-Matic* (vacuum advance only) type distributor used from 1949-1953. These are not suitable for performance applications. New point-type and electronic distributors are available as well as electronic conversion kits for all three types .

startup. The ballast resistor becomes part of the circuit after the starter is inactivated and slowly increases its resistance as current flows, eventually limiting current and voltage to the coil to about half the initial starting values. These early distributors are dual point, centrifugal advance (engine rpm dependent) with a vacuum brake to slow the advance when the throttle is suddenly applied. The vacuum brake is usually not needed with modern fuels, and the adjusting screw can be backed off to disable the brake. The early distributors required timing and advance characteristics to be set on a distributor test machine, unless the owner added a pointer and reference timing marks to the engine itself. **In general, these early distributors work fine even in performance applications if everything including the coil is up to specifications.** Conversion to 12 volts is just a matter of changing the coil, ballast resistor, and condenser, to the equivalent 12-volt items.

The distributors used from 1949 to 1953 however, are top mount units similar in appearance to more modern units used up till the mid 1980's on "modern" engines. Although similar in appearance to most modern distributors, the 1949-1953 stock flathead "Load-O-Matic" distributors are timed port (from carburetor venturi not manifold) **vacuum-only** advance units. Timing marks and a pointer are on the engine so the timing can be set with a conventional timing light. **These later stock distributors are adequate for stock engines, but are generally not suitable for performance engines with non-stock induction systems.**

AFTERMARKET IGNITION SYSTEMS

Modern new manufacture, breaker point or electronic ignitions are currently available for both the early and later flatheads. These units are generally centrifugal advance units that are fully adjustable for both initial and rpm dependent advance characteristics. The electronic units are generally 12-volt units that can be used with high output coils and various electronic "boxes" to tailor the spark characteristics of the engine to the application. In addition, there are modern true magneto and magneto look-alike electronic distributors available for the top mount units.

FLATHEAD AIR-FUEL TUNEUP

Determining optimum carburetor jetting or fuel injection parameters is a lot easier than it once was. The current widespread availability of cost effective air-fuel ratio meters allows anybody to tune any vehicle for top performance and economy without guessing or trying to read sparkplugs. I highly recommend units from Innovate, Fast and others. The flathead has air-fuel (A/F) ratio requirements similar to many modern engines. The ideal or stochiometric ratio for pump gasoline is 14.6 to 1, i.e. 14.6 pounds of air to 1 pound of fuel. In practice, the desired A/F ratio is approximately 14.0 to 15.0 for economical, part throttle cruising. For wide open throttle, peak power A/F ratios in the range of 12.0 to 13.0 give best results for normally aspirated engines. Supercharged engines can often produce peak power at ratios as rich as 10.5 because of the inter-cooling effect of the rich mixture. The case studies in the following chapters report the best A/F ratios for each application tested.

Most carburetors do an amazing job of maintaining the desired A/F ratio over broad operating conditions with only minimal changes, if properly sized for the application. I have been very impressed with the performance of the old standby Stromberg ***97***'s. They have a mechanically operated power valve which is more predictable in multi-carburetor applications than the equivalent Holley ***94***'s with their vacuum operated power valves.

FLATHEAD IGNITION TUNEUP

Determining optimum ignition timing is generally tougher than determining optimum air-fuel ratio because of all the variables involved, and the instrumentation required. As outlined in ***Blown Flathead***, the flathead is generally a fast burn engine and therefore does not require large advance numbers, and gives peak performance with relatively low peak timing numbers compared to most overhead valve engines. **Peak power for most normally aspirated flatheads is obtained with total advance numbers of 22-26 degrees.** Most supercharged engines like about 6 degrees less, because the cylinder pressures are much higher. Too much advance leads to rapid loss of power and ultimately engine failure if detonation occurs. Unless you are tuning on a dyno with a good knock sensor, be conservative!

Figure 5-8 is a stock flathead factory dyno curve that includes the full throttle ignition advance curve for the test engine. A part load curve (part throttle, high vacuum) would look similar to the full throttle curve, except more ignition advance would be required because the lower density air-fuel mixture burns more slowly. Saying this another way, as air-fuel mixture density

SERVICE BULLETIN

PAGE NO. 59 SUBJECT NO. S-6000

Wide Open Throttle Engine Performance Curves

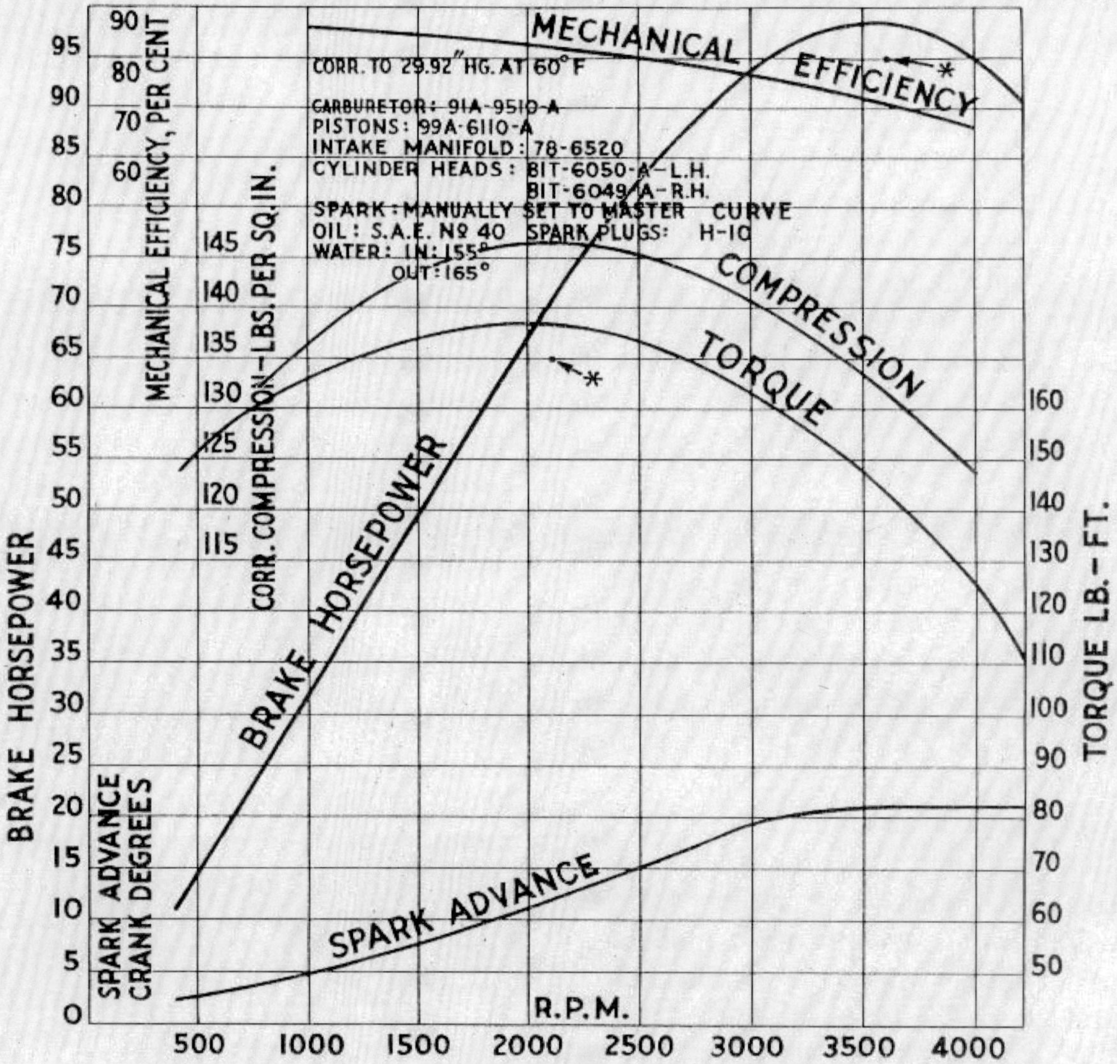

ABOVE TESTS MADE WITH GENERATOR AND WATER PUMPS BUT WITHOUT FAN OR MUFFLER.
* Advertised maximum torque and brake horsepower.

Fig. 1

95 H.P. STARTING 1939

Printed in U.S.A.

February 15, 1939

Figure 5-8. This is factory dyno sheet for a stock 239 cu.in. Ford flathead. Note the low and slow ignition advance curve. Peak horsepower occurred at 3600 rpm!

Intake Manifold	Carburetion	Exhaust Headers	Mufflers	Tailpipe(s)	HP Average 2000 - 4000 rpm	HP Peak @ rpm	A/F Ratio 2500 - 3500 rpm
Stock Single	Stock Holley 94 # 51 main jets	Stock cast with crossover pipe	Single Stock-type	Single Stock-type	82	90 @ 3400	
Stock Single	Stock Holley 94 # 51 main jets	Rear dump tubular	Duals Stock-type	Dual Stock-type	85	93 @ 3500	
Stock Single	Stock Holley 94 # 51 main jets	Center dump tubular	4" ID straight-thru duals	Duals 4" ID shorty	88	96 @ 3600	11.7 avg
Stock Single	Stock Holley 94 # 49 main jets	Center dump tubular	4" ID straight-thru duals	Duals 4" ID shorty	87	96 @ 3600	13.2 avg
Offenhauser Triple	Stock Stromberg 97 # 45 main jets	Center dump tubular	4" ID straight-thru duals	Duals 4" ID shorty	93	108 @ 3600	11.8 avg

Table 5-2. 59A Ford Flathead Exhaust and Intake Systems Evaluation. Basic engine is a stock 1946-48 *59A* Ford flathead, 239 cu.in., with stock water pumps and no fan or air cleaner(s). See photos in Figures 5-9 and 5-10. Horsepower (HP) values above are corrected to standard sea level dyno conditions (60 F, 29.92 in. Hg, dry air).

goes up (throttle open more and more load), less advance is required since the high-density mixture burns faster. Also, as engine speed increases, more advance is required to get the mixture lit in time to produce peak cylinder pressure when the piston is in the optimum position for peak power. All of this suggests that a lot of engine dyno time is required to produce 3-D "maps" of optimum ignition advance versus engine load and rpm. Luckily the flathead is happy over a relatively broad range of ignition timing and air-fuel mixtures, and staying in the middle of the recommended values is good enough in most cases.

EVALUATION OF INTAKE AND EXHAUST MODIFICATIONS ON A STOCK *59A* FLATHEAD

Dynamometer testing and analyses of a stock, second generation flathead Ford engine were conducted at Roadrunner Engineering to evaluate the stock engine and several "bolt-on" intake and exhaust systems. Measured average and peak horsepower values for the basic stock 239 cu.in. flathead in several naturally aspirated configurations are shown in Table 5-2.

Testing was conservatively limited to 4000 rpm since the HP peaks were all below 4000 rpm. Peak power for all tests systems occurred at approximately 3600 rpm, which is typical of an essentially stock flathead (see factory dyno curve in Figure 5-8). 91-octane premium pump gas was used for all tests. Testing was performed with a dummy generator and the stock water pumps installed. No fan or air cleaners were used. Three runs with each test configuration were averaged for the reported test results. All power values are corrected to standard dyno (sea level) conditions of 60 F and 29.92 in. Hg.

The engine was subjected to 35+ dyno runs ("pulls") over the test period. The first series of pulls on the unmodified engine was performed to provide a baseline for subsequent testing.

The second and third series of tests evaluated the effects of several dual exhaust systems on the otherwise unmodified engine. Most owners who add a dual exhaust system do it for the sound and the look. Nevertheless, there is a modest improvement in performance, fuel economy and operating temperature. The test engine gained about 3-6 hp (peak) depending on the exhaust system design as shown in Table 5-2. A free breathing, larger displacement engine would obviously benefit more.

The fourth series of tests was identical to the third series except the Holley ***94*** carburetor was re-jetted from #51 jets to #49 jets, to lean out the mixture. Very little difference in power was noted as shown in Table 5-2. This indicates that the optimum air-fuel mixture is probably somewhere between 11.7 and 13.2, or about 12.5.

The fifth series of tests evaluated a multiple carburetion system utilizing the best dual exhaust system (the center-dump system shown in Figure 5-10). Adding additional intake capacity is a proven horsepower adder and adds to the "coolness" factor. In this series of tests, a three-carburetor Offenhauser manifold was tested with a set of restored Stromberg ***97*** carburetors as part of a customer evaluation. The customer prepared all carburetors. Even though three carburetors on a stock flathead might be considered too many, the engine responded well with very little stumbling or hesitation even with the straight (non-progressive) linkage. A progressive linkage would make this a very practical setup on the street. These carburetors used stock #45 sea level main jets for the tests, which caused the full throttle mixture to be somewhat rich (11.8 A/F) at the 5300 ft test altitude. This is slightly richer than the generally accepted peak power optimum of 12.5, but previous testing demonstrated that A/F ratios in this ballpark do not cause a significant reduction in peak power. The triple carburetor setup provided a gain of about twelve additional hp over the single unit. See Table 5-2. The triple carburetor setup also demonstrated more uniform exhaust gas temperatures than the single setup indicating better fuel distribution.

Overall, the engine responded remarkably well to the bolt-on modifications, picking up 18 hp (+20%) with the best aftermarket intake and exhaust tested. This magnitude of increase would certainly be noticeable on the street.

EVALUATION OF INTAKE SYSTEMS AND IGNITION TIMING ON A MODIFIED *8BA* FLATHEAD

A variety of intake systems were evaluated on a mild performance, 262 cu.in. flathead long block assembly. This engine was mildly ported with a "3/4" race cam (.340" lift, 220 duration @ .05" lift), and fitted with a set of Offenhauser ***425*** heads (8.1:1 compression ratio on this engine), and a heavy duty main bearing support. The center dump dyno headers and mufflers were used for all tests. Figure 5-11 shows the engine on the dyno with a Carter ***WCFB*** (white metal cast four barrel) carburetor installed on an Offenhauser manifold. The test results for this setup and others are summarized in Table 5-3.

Figure 5-9. A stock *59A* flathead with a single Holley *94* carburetor and single exhaust shown on the Roadrunner Engineering dynamometer.

Figure 5-10. A stock *59A* flathead modified with 3x2 carburetion and center-dump headers. Although not shown here, mufflers and tail pipes were used during testing. The engine picked up 18 HP compared to the configuration shown in Figure 5-9.

Figure 5-11. Jim Glover's 262 cu.in. Flathead on the Dyno. The photo shows the engine with a Carter *WCFB* 4-bbl installed on an Offenhauser manifold. See results below.

Peak TQ	@ RPM	Peak HP	@ RPM	Total Adv	Main Jets	A/F (avg)
Naturally Aspirated, Stock Intake with Single Holley *94*						
226	2600	142	4000	26	0.049	13.0
Naturally Aspirated, Offy 4-bbl Manifold with Carter *WCFB* (385 CFM)						
232	3000	168	4500	26		12.7
Naturally Aspirated, Offenhauser *Triple* with Stromberg *97*'s						
236	2900	176	5000	26	0.043	12.5
Weiand *142* Supercharger (4 psi boost) with Holley 570 *Street Avenger*						
274	3900	249	5000	18	.054/.066	13.2
273	3900	251	5000	20		
279	3800	251	5000	22		

Notes:
1) All test results are an average of two or more tests (dyno "pulls").

Table 5-2. Jim Glover 262 cu.in. Flathead Dyno Test Summary

-6-

Supercharging the Flathead

Figure 6-1. An Italmeccanica/S.Co.T. Supercharged Flathead. A gorgeous engine combo! Besides the looks, the supercharger brings flathead performance up to modern levels. The S.Co.T. is currently reproduced by H&H and others.

SUMMARY

In this chapter we will take a general look at various flathead supercharging options. The what, why, and how of supercharging the flathead will be discussed along with examples of several systems. Key elements of this chapter are taken from ***Blown Flathead***.

Figure 6-2 shows the most common types of superchargers used in automotive applications. The terms supercharger and blower are used interchangeably in this book. All of the supercharger types in Figure 6-2 have their pros and cons, and all have been used in the

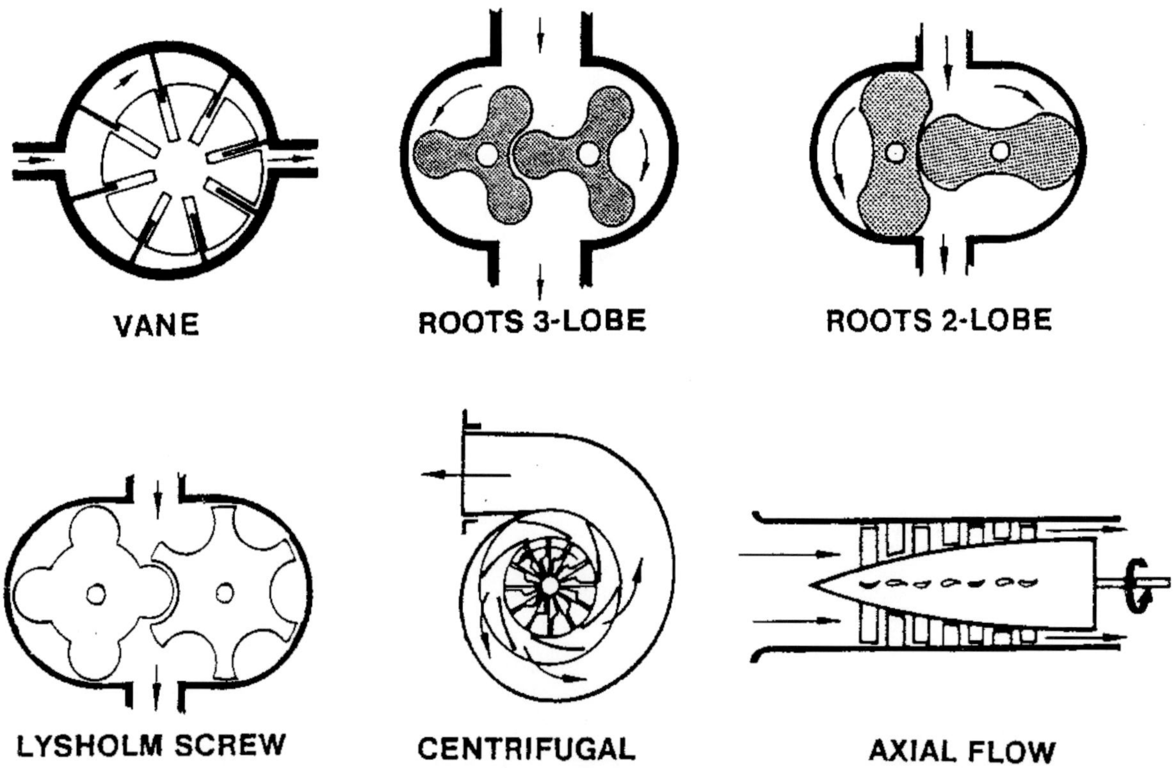

Figure 6-2. Types of Superchargers

past on the flathead. Currently, only the Roots type is offered commercially in kits for the flathead. The first commercial kits that I am aware of for the flathead were the centrifugal McCulloch units offered from the mid 1930's to the 1950's. Even though the centrifugal units worked well, the most popular kits for the flatheads have been Roots types since the late 1940's.

Turbochargers (exhaust driven superchargers) have been used on flatheads, but the fact that the exhaust is water cooled to a large extent on the flathead limits the performance gain from the turbo systems.

The reasons for the Roots type popularity include their reliability, moderate cost and great drivability (capable of predictable boost from off-idle to full throttle). The Lysholm screw type compressors are more efficient (at greater cost), but can be sized and analyzed similarly to the Roots types. This book will deal specifically with the Roots type superchargers, although much of the information is relevant to any supercharger.

WHY SUPERCHARGE?

The simple answer is that supercharging is the single most effective way to help a breathing challenged engine such as the flathead to make serious horsepower. In ***Blown Flathead,*** we went through the discussion of how supercharging raises the mean effective pressure (MEP) in the engine cylinder making the engine run like a much bigger displacement engine. Increasing the displacement, raising the cylinder pressure, and raising engine rpm, all can produce higher power. All engines are limited in what we can do in all three of these areas. The flathead is more limited than most. The biggest displacement flatheads are still small by modern V-8 standards. The maximum safe rpm is limited by the structural limits on the three main bearing supports for the rotating assembly. The intake ports also limit breathing and thus also limit rpm. And finally, the large surface area of the flathead combustion chamber limits the peak cylinder pressure we can endure before the head lifts. We have our work cut out for us to make flathead horsepower, but supercharging has proven to be the preferred path for the street. We just have to know where the limits are.

WHAT IS BOOST AND HOW DO WE MEASURE IT?

Supercharger "boost" is the increase in intake manifold pressure due to the action of the supercharger, normally quoted under full throttle conditions. The boost will vary with engine rpm depending on the characteristics of the engine and the supercharger. Anytime the supercharger is compressing more air than the engine can "inhale," the boost will show up as a positive pressure on a boost gage. Maximum boost at any rpm for a draw-through (e.g. carburetor at the supercharger inlet) system will occur at full throttle and will vary with rpm as stated previously. A typical boost gage is a differential pressure gage, i.e. it shows the difference between the measured pressure and atmospheric pressure, and gages are available that will indicate both negative (vacuum) and positive (boost) pressure. A normally aspirated engine will show a negative pressure (vacuum) in the manifold at all times, while a draw-through supercharged engine may indicate either vacuum or boost depending on the throttle position when measured with a common differential pressure gage.

WHERE DO WE START TO SUPERCHARGE OUR FLATHEAD?

The approach is as follows:

1. Read this chapter until you understand the limitations, pitfalls and advantages of supercharging. Particularly strive to understand and avoid detonation.
2. Estimate your compression ratio from Table 4-1, Flathead Compression Ratio (page 35).
3. Estimate your boost and octane requirements using Figure 6-4 (page 65).
4. Replace or modify the heads if necessary to be compatible with your choice of fuel octane.
5. Determine an appropriately sized supercharger based on the recommendations and Table 6-1 in this chapter (page 67).
6. Determine the pulley ratio required to meet your boost and performance requirements. You can use kit manufacturer's recommendations (preferred) or estimate yourself based on analysis and/or techniques in this chapter (page 68).
7. Determine the availability and cost of kits and/or subassemblies that meet your requirements.
8. Install the unit per manufacturer's instructions (if applicable) and follow tuning recommendations in this book.
9. Have an overhead engine powered car for lunch and be kind to their owner!

PRE-IGNITION AND DETONATION

Pre-ignition and detonation are serious threats to any engine and especially supercharged engines. Under pre-ignition or detonation conditions, cylinder pressure can easily be many times the normal cylinder pressure, usually causing rapid engine failure.

Pre-ignition is an early and usually unintended ignition of the air-fuel mixture. Firing the mixture too early in the power cycle results in wasted energy and potentially extreme structural loads on the piston and other rotating parts. Pre-ignition is often the result of hot spots in the combustion chamber, such as the spark plug electrode, other **sharp edges**, the exhaust valve, or areas near any air or steam pockets in the head cooling passages. Excessive intake mixture temperatures and pressures also contribute to pre-ignition.

Detonation is an extremely rapid, **uncontrolled** burning (actually an explosion) of the air-fuel mixture, which may occur at multiple sites in the combustion chamber either before or after normal ignition. The results and causes are often identical to those of pre-ignition, such as hot spots and excessive temperatures and pressures. Detonation can also be caused by many other factors, such as early ignition timing (too much ignition advance before top dead center), insufficient fuel octane, and others. Excessive temperature and pressure in turn result from over-boosting (over-supercharging), too high a compression ratio, an inefficient supercharger (overheats the air-fuel mixture), and high intake air temperatures (e.g. under the hood on a hot day). Actual fuel octane of the air-fuel mixture can be degraded by excessive oil introduced by worn rings, valve seals or blower oil seals.

The symptoms for both pre-ignition and detonation include low power, overheating, "running on" after the key is turned off, audible "pinging" or "knocking," blown (that word again) head gaskets or total destruction of the engine. Pinging and lowered power are usually the first symptoms of pre-ignition or detonation. Pay attention!

Figure 6-3 shows detonation damage to a normally aspirated flathead caused by a combination of high compression ratio (9.7:1), too much ignition advance, and low octane fuel. Excessive oil burning also contributed (oil lowers the effective octane of

Figure 6-3. Detonation Damage.

the fuel). The detonation occurred at the furthest distance from the spark plug (the extreme left in the photo), which is typical. See Appendix D for why this is so. The engine still ran even though pieces of the piston had broken off and were hitting the head (marks on the top half of piston in the photo).

PREVENTING PRE-IGNITION AND DETONATION

Ignition timing, air-fuel ratio, engine compression ratio, supercharger boost pressure, and fuel octane are the major variables affecting pre-ignition and detonation, which we can control. Timing and air-fuel ratio will be covered under tuning. For now we will assume that timing and the air-fuel ratio are optimized. We are interested in engine compression ratio because it is a very important variable affecting peak cycle temperatures and pressures. Induction pressure (atmospheric plus supercharger boost pressure) and temperature are likewise important for the same reason.

As a very general guide, for sea level, low humidity, 77 degree F blower inlet air, the minimum octane required for gasoline to prevent pre-ignition and detonation in our flathead with different compression ratios is shown in Figure 6-4. Compression Ratio (CR) is the actual or estimated volumetric compression ratio of the engine. See Table 4-1. Pump gasoline octane is the average of Research (R) and Motor (M) octane, i.e. (R+M)/2. This is the number usually posted at the pump.

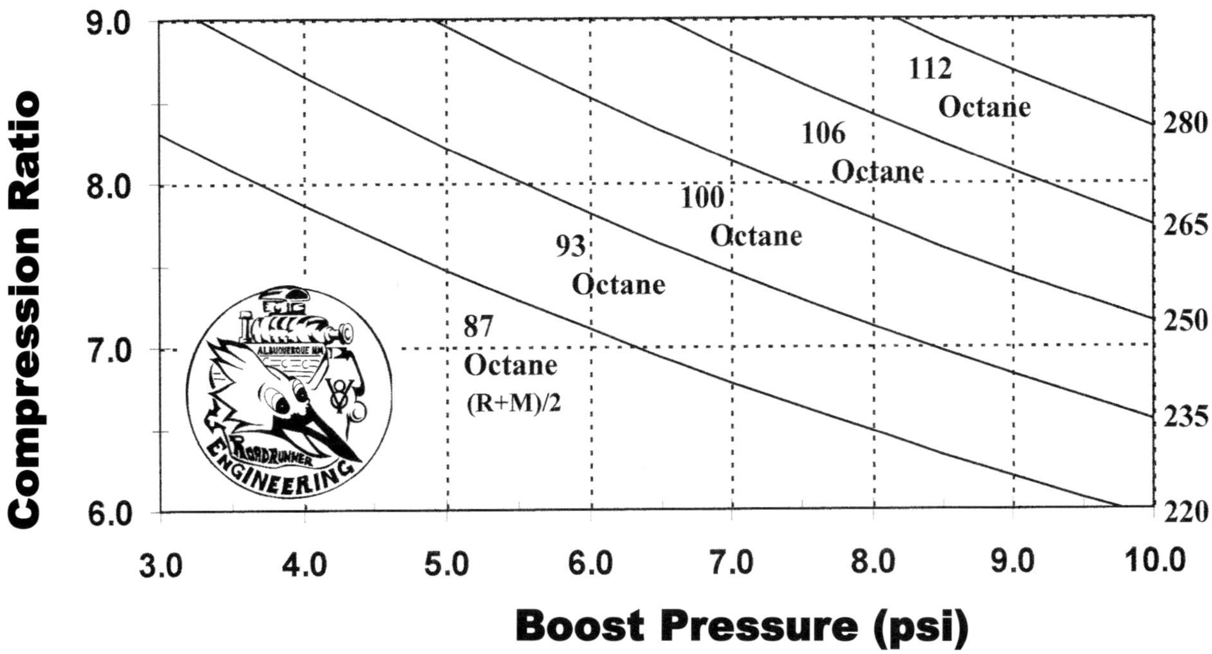

Figure 6-4. This plot shows octane requirements vs. safe operating supercharger boost pressure and engine compression ratio for the flathead. Compression ratio is estimated by using Table 4-1 in Chapter 4. Estimated boost is determined by the supercharger kit supplier based on experience and sophisticated computer models. Simple estimating techniques are also described in this chapter. Numbers at the right of the graph correspond to approximate horsepower for a 276 cu. in. engine.

As the humidity increases, the octane required to prevent detonation is reduced. If atmospheric pressure is lower (higher altitudes), it also reduces boost and the octane required. Albuquerque and Denver are about a mile above sea level and atmospheric pressure is about 12.4 psi or about 15% lower than at sea level. However, due to the higher compression and compression heating required to get the boost pressures in Figure 6-4, the octane required is about the same at higher altitude as for sea level for the same boost. If the blower inlet air temperature is greater than 77 degrees F (it could be 177 degrees F under the hood in the summer), we must reduce the timing, reduce the maximum boost levels, or increase the fuel octane.

BOOST AND COMPRESSION RATIO VS. POWER

Note the horsepower figures associated with the octane lines in Figure 6-4. The primary determining factors for octane requirements are the pressure and temperature in the combustion chamber. These same cylinder pressures and temperatures, along with the engine displacement are also the primary factors in determining the power output of the engine. The horsepower numbers at the right side margin of Figure 6-4 are estimates for a 276 cu.in. engine.

We note that for the street, about 6 to 7 psi (lb/in^2) boost with an 8:1 compression ratio (CR) engine requires 100-octane gasoline at sea level. 100-octane aviation or race gas costs about 50-100% more than 93-octane premium automobile gas. As an alternative we can use gas additives, water injection or rich air-fuel mixtures to get the required equivalent octane by trial and error. We can also use boost retard devices (described later). **However, limiting ourselves to 5 to 6 psi boost and a 7.2:1 to 8.2:1 CR allows us to use premium pump gas and still safely get about a 40% horsepower boost over and above whatever the engine was producing normally aspirated.**

Although not important for a street engine, alcohol can tolerate high boost pressures and/or high compression ratios even though its octane rating is only about 100, because it has a much greater cooling effect than gasoline.

For racing, where fuel economy and fuel cost are not important considerations, maximum power with minimum risk can be obtained with very high boosts and very low compression ratios. Typical values for alcohol-nitromethane fuel burning engines might be 20+ psi boost with 5.5:1 compression ratios! The high boosts and low compression ratios for these engines promote high volumetric efficiency and high mean effective pressure (MEP) while minimizing the destructive maximum pressure "spikes" associated with higher compression ratios discussed earlier.

RECOMMENDED BOOST AND COMPRESSION RATIO

Based on the considerations in the previous section, **I recommend 5 to 6 psi maximum boost levels with 7.2:1 to 8.2:1 compression ratios as practical combinations for everyday driving at sea level.** This allows the use of readily available unleaded premium fuel of 93 octane and the engine can be essentially stock if everything is up to specifications. For weekend drag strip duty, higher boosts up to 10 psi or more with a 7.5:1 CR delivers excellent performance, but requires racing gas, forged pistons and at least minimal center main bearing support (e.g. a "strap").

The boost, of course, can be tailored by the size of the blower, and the blower and crankshaft pulley combination. Thus, if the rest of the engine is built to take the high boost, you can run more aggressive pulley combinations on the weekend. It should be noted that boost pressure is not only a function of atmospheric conditions and the blower capacity, but also the breathing capacity of the engine. A "free breathing" engine will "inhale" and run better while indicating a lower boost pressure in the manifold than a more restricted engine, all other things being equal. It is not unusual for the full throttle boost pressure to increase by a factor of two or so going from 2500 to 5000 rpm with a positive displacement supercharger. A lower increase in boost pressure, again indicates a better breathing engine as noted above. Breathing is influenced by the air cleaner, carburetion, blower characteristics, intake and exhaust manifold design, cam and valve train, and head and block porting parameters. Although boosting the flathead with a supercharger helps overcome it's basic breathing restrictions, **better breathing means better performance, blown or unblown.**

BLOWER SIZE FOR SUPERCHARGED ENGINES

There is an optimum supercharger capacity (size) range for each engine that will provide good overall performance and throttle response. Too small and the engine will not perform up to it's potential. Too large and throttle response and fuel economy suffer.

Roots-type blower capacities of about one-half to two-thirds the engine displacement per revolution work very well for street and weekend drag strip duty. See Table 6-1 for blower capacities. Smaller capacity units require higher rotational speeds to increase capacity with a corresponding decrease in blower life. Bigger units tend to be inefficient due to rotor leakage losses at lower rpms with high boost, but may be favored for sustained high speed driving or appearance. **An oversized supercharger that is under-driven**

Blower/Supercharger Make & Model	Capacity (Cu.In./Rev)	Rotor Design
GMC 2-71	136	3 lobe twisted
GMC 3-71	210	3 lobe twisted
GMC 4-71	280	3 lobe twisted
GMC 6-71	420	3 lobe twisted
GMC 4-53	139	2 lobe straight
GMC 6V-53	208	2 lobe straight
GMC 8V-53	334	2 lobe twisted
Italmeccanica/SCoT (8.63" long rotors)	172	2 lobe straight
Canden 7"	112	2 lobe straight
Canden 9"	144	w/ Teflon tips
B&M/Holley 144	144	2 lobe straight w/ Teflon tips
B&M/Holley 174	174	
B&M/Holley 250	250	
Eaton M112	112	3 or 4 lobe twisted
Weiand Pro-Street 142	142	2 lobe straight
Weiand Pro-Street 177	177	2 lobe straight
Weiand Pro-Street 256	256	2 lobe straight

Table 6-1. Air moving capacity (displacement per revolution) and rotor design of selected Roots-type blowers/superchargers.

sufficiently to prevent over-boosting at high speeds, may actually "get in the way" at lower engine speeds! This often results in severe drivability problems. Bigger is not always better!

The proper choice of the blower and crank pulley sizes are of course dictated by our performance goals, atmospheric pressure, the size (displacement) of the engine and its breathing characteristics, the size of the blower and its characteristics, and other factors such as fuel cost, etc. The blower kit suppliers should help here. There is no "one size fits all" in the supercharger world. **Buyer beware!**

ESTIMATING BOOST AND BLOWER PULLEY RATIO FOR SUPERCHARGED FLATHEADS

There is no simple, accurate relationship for estimating boost and pulley size because of all the variables involved as outlined in the previous section. To do it best requires a sophisticated analysis for your specific application. Where a more sophisticated analysis is not available, I use an approximate approach for doing preliminary pulley sizing. **In any case, we must first determine our compression ratio and then pick a safe boost level for the fuel octane we choose.**

As an example, if we measure our head combustion chamber volume to be 76 cc and consult Table 4-1 for a relieved 276 cu.in. engine, we find that our compression ratio is 7.5:1. Then consulting Figure 6-4, we decide that we want to use 93-octane pump gas. This limits the boost to approximately 6 psi for safe operation. Next we are ready to pick a blower size and size the pulleys. From the blower size discussion previously, we know that a typical 276 cu.in. engine will perform well with a Roots blower displacement of 138-185

Determining Supercharger Drive Ratios
Method A– Huntington Equation

A general approach for estimating the supercharger drive ratio was outlined by Roger Huntington in his 1950 book, ***Souping the Stock Engine.*** A modified version of the equation he used for estimating is given below.

$$R = \frac{(A + B) \times D}{A \times K \times V} \quad \text{where,}$$

R = blower drive ratio = crankshaft pulley diameter ÷ blower pulley diameter
= pulley ratio (or over/under drive)

A = atmospheric pressure in pounds/square inch (psi), which varies with altitude
= 14.7 psi at sea level (reduce by 3% per 1000 feet of elevation above sea level)

B = boost pressure in psi

D = engine displacement in cu.in.

V = Roots blower displacement in cu.in. per revolution

K = a constant determined by experience. A good starting point is 1.7 for a high performance ("3/4 race") street flathead with a 142 cu.in. displacement blower. Huntington used 1.6 as the constant.

cu.in. per revolution. For purposes of illustration, we will pick a Weiand ***142*** blower. Now we are ready to estimate the pulley drive ratio.

Method A — Equation

Continuing our example, assume we have a 276 cu.in. engine (D) with a 142 cu.in. per revolution blower (V), and we want 6 psi boost at sea level (B).

Now, we can calculate the drive ratio using the equation in the yellow box,

$$R = \frac{(14.7 + 6) \times 276}{(14.7 \times 1.7 \times 142)} = 1.61$$

Next, we would look for or make a pulley combination to give the desired drive ratio of 1.61.

Method B — Tables 6-2 and 6-3

Evaluating the Huntington Equation for two common blower displacements yields the blower drive ratios listed in Tables 6-2 and 6-3. Table 6-2 lists estimated blower drive ratios for common engine displacements for a Roots street blower with a displacement of approximately **142 cu.in. per revolution.** The 142 size blowers include the GMC ***4-53***, the Weiand ***142***, the B&M ***144*** and the Camden 9 inch units. Table 6-3 lists estimated blower drive ratios for the same engine displacements for a Roots street blower with a displacement of approximately **174 cu.in. per revolution.** The 174 size blowers include the Weiand ***177***, the B&M ***174*** and the Ital-meccanica/S.Co.T. ***172***.

INTERCOOLING SUPERCHARGED ENGINES

An intercooler is a heat exchanger used to cool the compressed air mixture leaving the supercharger. This provides a cooler, denser mixture to the engine, resulting in higher power output and reduced detonation tendencies. However, these benefits come with a substantial penalty in cost, weight, and complexity. Also, a drop in boost pressure will be noted due to the cooling of the compressed air-fuel mixture and also due to flow resistance in the intercooler itself. The gains are usually not worth the cost and ef-

Engine Displacement in Cubic Inches	239	255	276	284	296	304
Blower Drive Ratio for 5 psi Boost	1.33	1.42	1.53	1.58	1.64	1.69
Blower Drive Ratio for 6 psi Boost	1.39	1.49	1.61	1.66	1.73	1.77

Table 6-2. Blower Drive Ratios vs. Engine Displacement for a 142 cu.in./revolution Displacement Blower. The boost numbers shown above are for full throttle at 4500 rpm engine speed at sea level with a 142 cu.in. per revolution capacity blower.

For our 276 cu.in. flathead example, we see the 1.61 ratio that provides our desired 6 psi boost as determined by Method A. The number for this example is shown in red.

Engine Displacement in Cubic Inches	239	255	276	284	296	304
Blower Drive Ratio for 5 psi Boost	1.08	1.16	1.25	1.29	1.34	1.38
Blower Drive Ratio for 6 psi Boost	1.14	1.21	1.31	1.35	1.41	1.45

Table 6-3. Drive Ratios For a 174 cu.in./revolution Blower. The boost numbers shown above are for full throttle at 4500 rpm engine speed at sea level with a 174 cu.in. per revolution capacity blower.

fort involved for street engines running boost levels of about 6 psi or less. Competition engines running boost levels of about 8 psi or more for sustained periods may justify or even require the use of an intercooler.

BYPASS VALVES FOR SUPERCHARGED ENGINES

A bypass valve is used on "blow-through" positive displacement supercharger systems. They are necessary to prevent the air inlet ducting between the blower and the throttle valve from being overpressured when the throttle is closed at high engine rpm when boost is greatest. This valve also reduces air heating and helps fuel economy under part throttle cruising with these systems since the blower is not forced to pump against a high back pressure. **A "suck-through" supercharger system where the carburetor or throttle body is located at the supercharger inlet (such as the Roadrunner system) <u>do not</u> require a bypass valve.** Fuel economy benefits from a bypass valve are minimal with a suck-through system since the supercharger is generally pumping in a partial vacuum.

BLOW-OFF VALVES FOR SUPERCHARGED ENGINES

Positive displacement supercharger systems with a positive drive such as a chain, cog belt or gear drive require a blow-off valve (e.g. burst plate) to protect the system in the event of a backfire or other system overpressure. Systems such as the Roadrunner with a non-positive belt drive do not require a blow-off valve because they can slip (and backup) under extreme pressure conditions. Enclosed supercharged marine engines are not permitted by law (i.e. it's illegal) to use a blow-off valve because it releases a combustible mixture into the vessel.

SUPERCHARGED FLATHEAD AIR-FUEL TUNEUP

As noted in the last chapter, the flathead has air/fuel (A/F) ratio requirements similar to most modern engines. Also noted, the ideal or stochiometric ratio for pump gasoline is about 14.6 to 1, i.e. 14.6 pounds of air to 1 pound of fuel. In practice, the desired A/F ratio is approximately 14.0 to 15.0 for part-throttle cruising. For wide open throttle, peak power A/F ratios in the range of 11.5 to 12.5 give good results for supercharged engines. Peak power supercharged may be obtained at ratios as rich as 10.5 because of the inter-cooling effect of the rich mixture. In general, **peak power** for supercharged engines may occur with A/F ratios about 0.5 to 1.5 points lower (richer) than for a normally aspirated engine. Richer is also safer because rich mixtures delay the onset of detonation because of the cooling effect. If we use a four-barrel carburetor, this can be accomplished nicely by installing secondary jets 2 to 4 sizes larger than the primary jets. Running leaner on the primary jets prevents plug fouling and enhances part throttle fuel economy.

SUPERCHARGED FLATHEAD IGNITION TUNEUP

Again, as noted in the last chapter, determining optimum ignition timing is generally tougher than determining optimum air-fuel ratio because of all the variables involved and the instrumentation required. As outlined previously, the flathead is generally a fast burn engine and therefore gives peak performance with relatively low peak timing numbers compared to most overhead valve engines. **Peak power for most supercharged flatheads can be obtained with total advance numbers of 16-20 degrees.** It should be noted that when supercharged engines are operating at part load (less than full throttle), the vacuum/ boost gage often indicates a vacuum as if the engine were normally aspirated. More ignition advance under these part load conditions gives better fuel economy as with a normally aspirated engine because the lower density air-fuel mixture in the cylinder burns more slowly. Stating the reverse, as air-fuel mixture density goes up (throttle opened more, more boost), less ignition advance is required because the high-density mixture naturally burns faster. In most cases, as engine speed increases more advance is required to light the mixture in

Figure 6-5. Mallory (electronic or points-type) Distributor and MSD Boost Timing Master, P/N 5462 or 6462. This combination is recommended for any supercharged flathead. The boost timing master can also be used with a front mount distributor.

time to produce peak cylinder pressure when the piston is in the optimum position for peak power. As always, too much advance leads to rapid loss of power and ultimately engine failure if detonation occurs.

A recommended strategy for ignition timing for supercharged flathead engines is as follows. Use a centrifugal advance distributor with an initial timing of 0-4 degrees at engine idle and a total timing of 24-26 degrees at cruising rpm, <u>combined</u> with a boost retard device that retards timing approximately 1 to 1-1/2 degrees per psi of boost. MSD boost retard modules (P/N 5462 or 6462) and others, can provide automatic electronic boost retard for any ignition, either points or electronic. These modules are recommended for all supercharged engines. Unless you are tuning on a dyno with a good knock sensor, be conservative! Fortunately, the flathead is happy over a relatively broad range of ignition timing and air-fuel mixtures. Staying in the middle of the recommended values is good enough in most cases.

FLATHEAD BLOWER KITS

The following photos show several flathead blower kits which are, or were, commercially available.

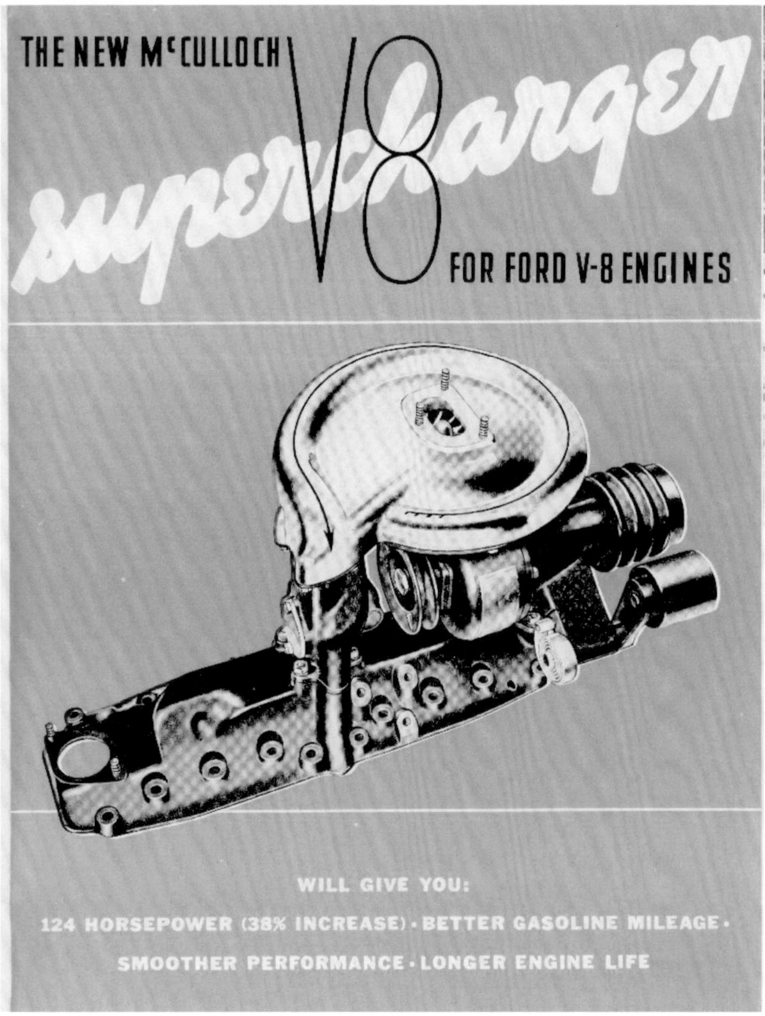

Figure 6-6. The McCulloch was the first commercially available kit for the Ford flathead. It was a centrifugal blower made in Milwaukee, WI and 1936 was the first year of manufacture. McCulloch claimed a 38% horsepower increase and up to 19.7% improvement in gas mileage. The cost was $124.50 and included a dual exhaust system.

Figure 6-7. Navarro 4-71 GMC Blower Installation. This kit is available from H&H. Other GMC-based kits are available from Mitchell and others. Rodders have used GMC-based, home-built installations for years, using both the 53 and 71 series blowers.

Figure 6-8. B&M Flathead Supercharger Kit with the B&M *142 ProCharger* Installed. This picture is from their 1990 catalog. Later, this kit was offered by Motor City Flathead with a B&M *174 Pro-Charger* installed until 1998 when B&M's blower line was acquired by Holley Performance Products. Many of these kits are still on the road. The *142* unit pictured above and those that follow are generally similar to the 4-53 GMC blowers.

Figure 6-9. Camden Industries Flathead Supercharger and Manifold with a Prototype Roadrunner Drive. This unit was offered in both a 7" and 9" (rotor length) version in the 90's. The blower design itself was produced from 1957 until Camden ceased selling superchargers in the mid-1990's. This is a 9" unit purchased in 1994 and ran on the *Motorhead Mart Special* for several years. Speedway Motors also sold this unit under the "All American" name. Note the external oil lines for the plain sleeve-type blower bearings.

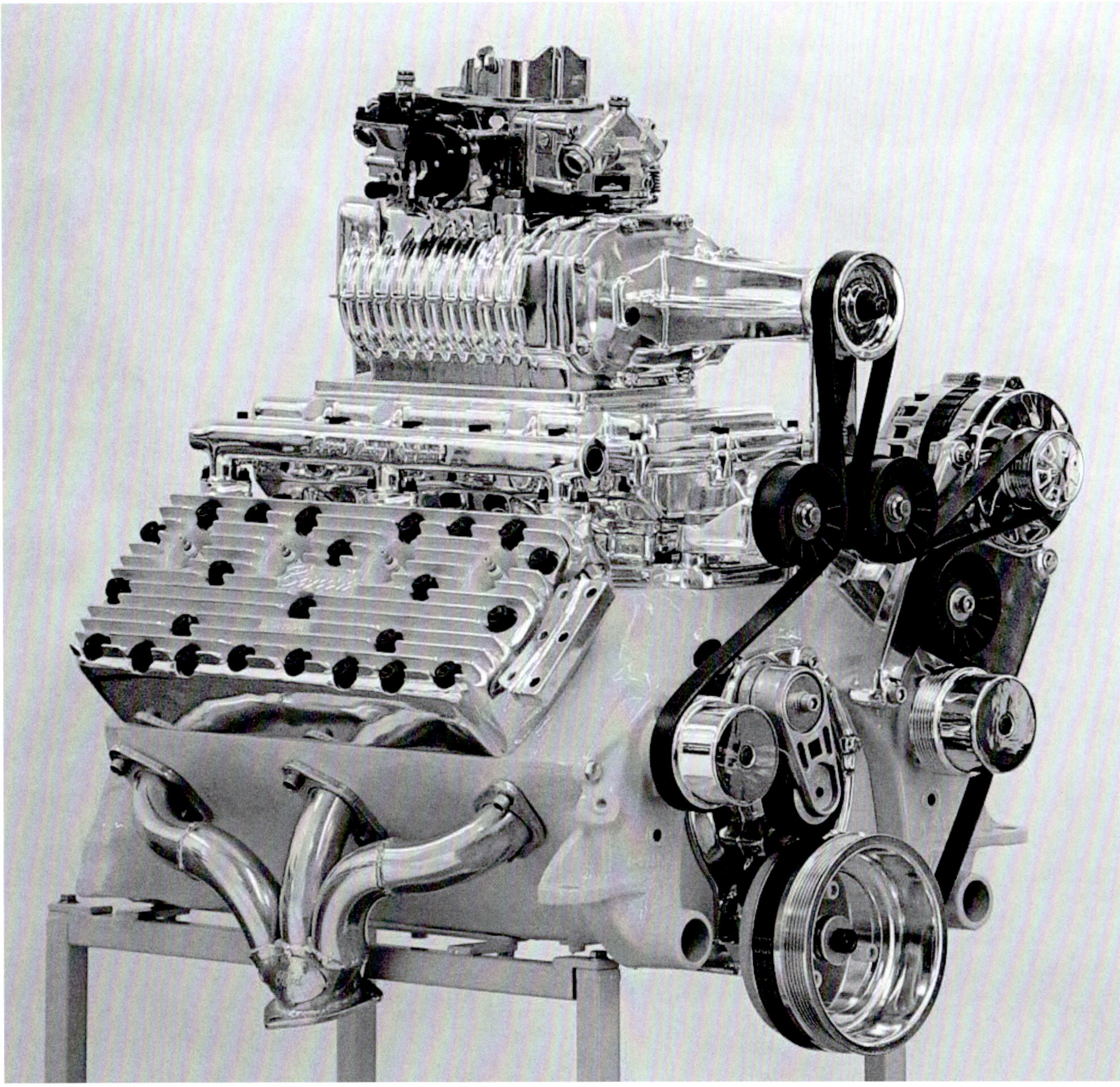

Figure 6-10. Magnuson/Roberts Supercharger Kit for the Flathead. This kit uses an Eaton *112*-based supercharger similar to those used on some Ford *Lightnings* and other OEM applications.

Figure 6-11. Roadrunner Engineering Street Kit with a Weiand *142* Supercharger (now produced by Holley) Installed. Roadrunner kits are available for both the *59A* and *8BA* flatheads and with either the *142* or *174* Weiand superchargers. The *174* unit is almost identical to the *142*, except that it is approximately 1.8" longer. The *174* still fits in front of the breather assembly and therefore does not affect firewall clearance.

-7-

John Downs 255 cu.in. Flathead Ford

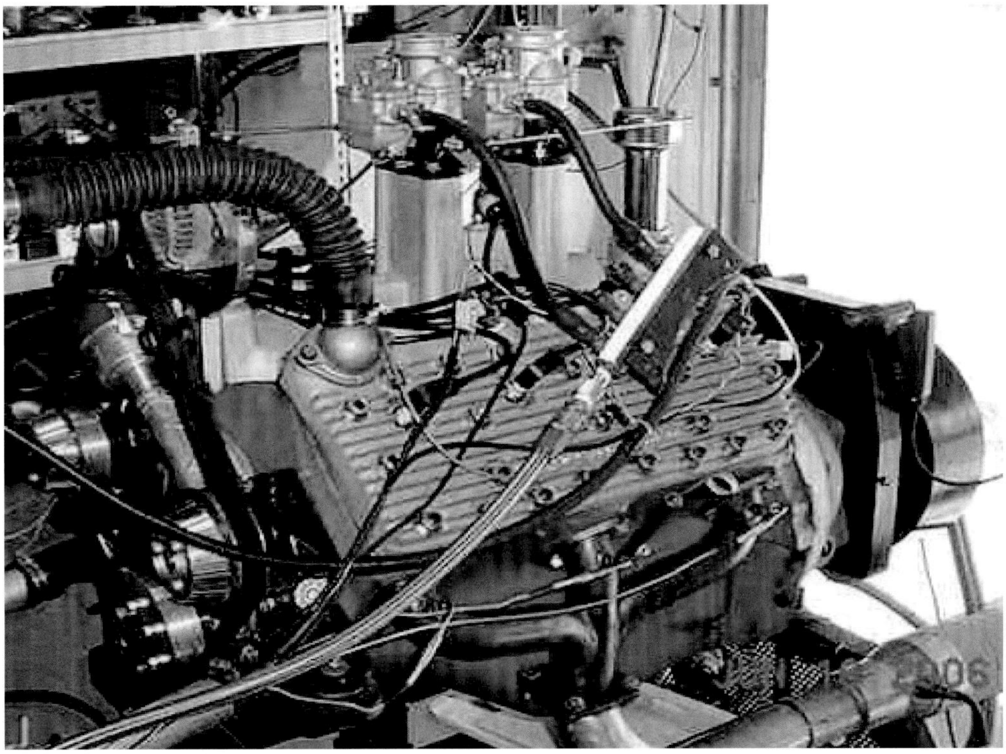

Figure 7-1. Downs 255 cu.in. Flathead During Normally Aspirated Testing on the Dyno. An original Thickstun *PM-7* high-rise manifold with Stromberg *97* carburetors is shown here.

SUMMARY

This engine was built on a tight budget and was originally intended to be a mild normally aspirated street engine. As a result, this engine was not equipped for maximizing power output either normally aspirated or supercharged. Nevertheless, this engine produced 176 hp naturally aspirated and 217 hp blown, all at a moderate 4500 rpm. Test rpm was conservatively limited to 4500 rpm, although 5000 rpm would probably have been fine even with the cast pistons and no special engine prep. The torque averaged over 255 ft-lbs from

2000 to 4500 rpm. Peak power for all induction systems occurred at the 4500 maximum test rpm indicating a free breathing engine that had much more potential. 91-octane premium pump gas was used for all tests. Testing was performed with an alternator and the water pumps installed.

TEST DESCRIPTION AND RESULTS

The engine was subjected to 40+ dyno runs ("pulls") over the period of 6/9-6/26, 2006. The first series of pulls was performed normally aspirated to break-in the engine, do initial tuning, and check for proper function of all components. Problems were encountered with water leakage around some head studs and a magnetic pickup module in the Mallory distributor failed during initial testing. The heads were removed and all studs resealed with "Leak Lock" before reassembly and no further leakage was observed. The heads were re-torqued after running to operating temperature and cooling for a total of five times per the Roadrunner kit instructions to assure proper head seal during supercharger boost. The majority of testing was conducted with a point type Mallory distributor until a replacement module (Mallory P/N 609) for the electronic version was obtained. Final tests had the electronic distributor re-

Figure 7-2. Downs 255 cu.in. Flathead During Normally Aspirated Testing. An Offenhauser *Triple* manifold and Stromberg *97* carburetors were used.

Figure 7-3. The Downs Engine with a Roadrunner Supercharger Kit on the Dyno. A Holley 570 cfm *Street Avenger* four barrel carburetor provided the intake mixture.

installed and functioning properly.

After initial "bugs" were fixed, normally aspirated testing resumed. Three intake manifolds with Stromberg ***97*** carburetors prepared by Jere Jobe of Vintage Carburetion Technology Inc. were used for these tests. The Thickstun ***PM-7*** high rise two-carb manifold, the Edelbrock ***Super Dual***, and the Offenhauser ***Triple*** manifolds produced averaged peak power readings of 138, 157 and 176 horsepower respectively, as shown in Figure 7-5.

The Edelbrock 2x2 outperformed the high rise 2x2 Thickstun manifold across the test rpm range (2000-4500 rpm). An unknown part of the difference is due to the heat riser, which was functional on the Thickstun. The Edelbrock and Offenhauser risers were blocked. Any benefit of the tall manifold would be expected at 3500 rpm or less due to the long runners. None was observed. The three-carb Offy manifold was the clear champ of the three normally aspirated candidates. No stumbling or hesitation was observed even with the straight (non-progressive) linkage. The carburetors used stock #45 main jets for most tests, which caused the full throttle mixture to be ideal (12.5:1) at the 5300 ft test altitude. #43 jets were also tried but had no significant effect on power. However, the spark plugs stayed cleaner with the smaller jets, and fuel mileage would be better at high altitude. Peak power normally aspirated was observed with total timing set at 25-27 degrees, although there was little difference over the range of 20-30 degrees.

Supercharged, the engine produced 217 hp and a remarkably constant 255 ft-lbs. of torque with only 3 psi boost. The high compression ratio (8.9:1) and fuel octane limited boost. A higher boost level would be desirable if the engine were to be operated at high altitude, rather than at low altitude like Phoenix.

For the supercharged testing, peak power was observed at 20-25 degrees total advance with virtually no difference in output over this range. Running on the low side of this range has the advantage of reduced detonation tendencies and lower octane requirements for the street. Jetting was not varied from stock for the Holley ***Street Avenger*** because the air-fuel mixture was close to optimum (11.5:1) for peak supercharged power at the test altitude. The very rich mixture gives an intercooling effect and reduces detonation tendencies. The ***Street Avenger*** is a vacuum operated secondary unit and is designed for easy secondary spring replacement (2 screws). The spring controls secondary opening characteristics, which can have substantial impact on throttle response and drivability. In addition to the installed spring, the ***Street Avenger*** comes with a lighter and heavier spring for adjusting to driver needs. The light spring produced a "bog" and instability in dyno operation at less than 2000 rpm, but might work fine at lower altitude on the street. Driver experimentation required.

Overall, this little engine performed remarkably well with all of the induction systems, and demonstrates what can be done on a budget without exotic parts. Future tests on other engines will include evaluation of "blower calibrated" carburetors, which pick up the vacuum signal for power valve operation from an external line to the intake manifold rather than at the base of the carburetor. The manifold vacuum/boost source should yield better throttle and power valve response. At full throttle, the power valve is wide open in any case.

Measured torque and horsepower curves for the mild 255 cu.in. flathead in both naturally aspirated and supercharged configurations are shown in Figure 7-5 at the end of this chapter. All power values are corrected to standard dyno (sea level) conditions of 60 F and 29.92 in. Hg.

The engine build information is listed below.

ENGINE DESCRIPTON AND SPECIFICATIONS

<u>Short block assembly</u>

a. 3-3/16" +.100" bore, cast pistons with stock dome.
b. 3-3/4" stroke Ford crank; stock rods, no main cap support.
c. 1.5" stainless intake and exhaust valves, block "pocket ported" and relieved.
d. Melling ***M19*** stock volume oil pump.
e. Isky ***Max1*** cam equivalent with single Isky ***185G*** valve springs, shimmed as necessary to get proper installed height. Valve clearance was 0.012" (intake) and 0.014" (exhaust).

<u>Heads</u>

Offenhauser ***400*** aluminum heads with an 8.9:1 compression ratio (54 cc head volume). This is high for a supercharged flathead but will work with moderate boost, premium pump gas, and a conservative advance curve and/or a boost retard device.

<u>Intake-Naturally Aspirated</u>

Three manifolds were evaluated as noted previously. Three stock Stromberg ***97*** carburetors were available for installation as required for each manifold.

Intake-Supercharged

A Roadrunner single-plane manifold and a Weiand ***142*** blower with 37% overdrive (5.625" crankshaft pulley, 4.10" supercharger pulley) was used for the supercharged tests. A 570 cfm Holley ***Street Avenger*** (4-bbl) carburetor with vacuum secondaries and stock jets was used. The three vacuum secondary springs furnished with this carburetor were tried and produced different low-end torque characteristics, but identical peak power readings. No air cleaner was used.

Exhaust

Free flowing center-dump headers and large truck mufflers were used during test. The engine had no center exhaust baffles.

Ignition

A top mount electronic (magnetic pickup) Mallory distributor with mechanical advance and high performance coil were used. Autolite ***4092*** sparkplugs gapped at .025"-.030" were also used. The total advance curve was limited to 20-25 degrees at about 2600 rpm for most of the supercharged runs. An ignition retard device such as those available from MSD (P/N's 5462 or 6462) with a retard setting of 1-2 degrees per psi of boost is recommended in service. Then the total timing can be set to 24-28 degrees for part load cruising economy yet prevent full throttle detonation.

Water pumps

Stock ***8BA*** type new pumps with modern seals and bearings from Speedway Motors were used along with Roadrunner serpentine pulleys and hubs for all tests.

PROJECT PARTICIPANTS

Automotive Machine Service in Albuquerque, NM built the subject engine for the owner, John Downs of Phoenix, AZ. Wayne Woody, Albuquerque, NM, did the porting. Testing and tuning were performed by Gary McGlasson of McGlasson Racing Engines, Los Lunas, NM. The author was the Project Manager.

Figure 7-4. A "ports eye view" of the Downs engine shows the stock intake ports. The valve "bowl" area under the valve was opened up, but is only partly visible here. The block was relieved between the valves and the cylinder bore.

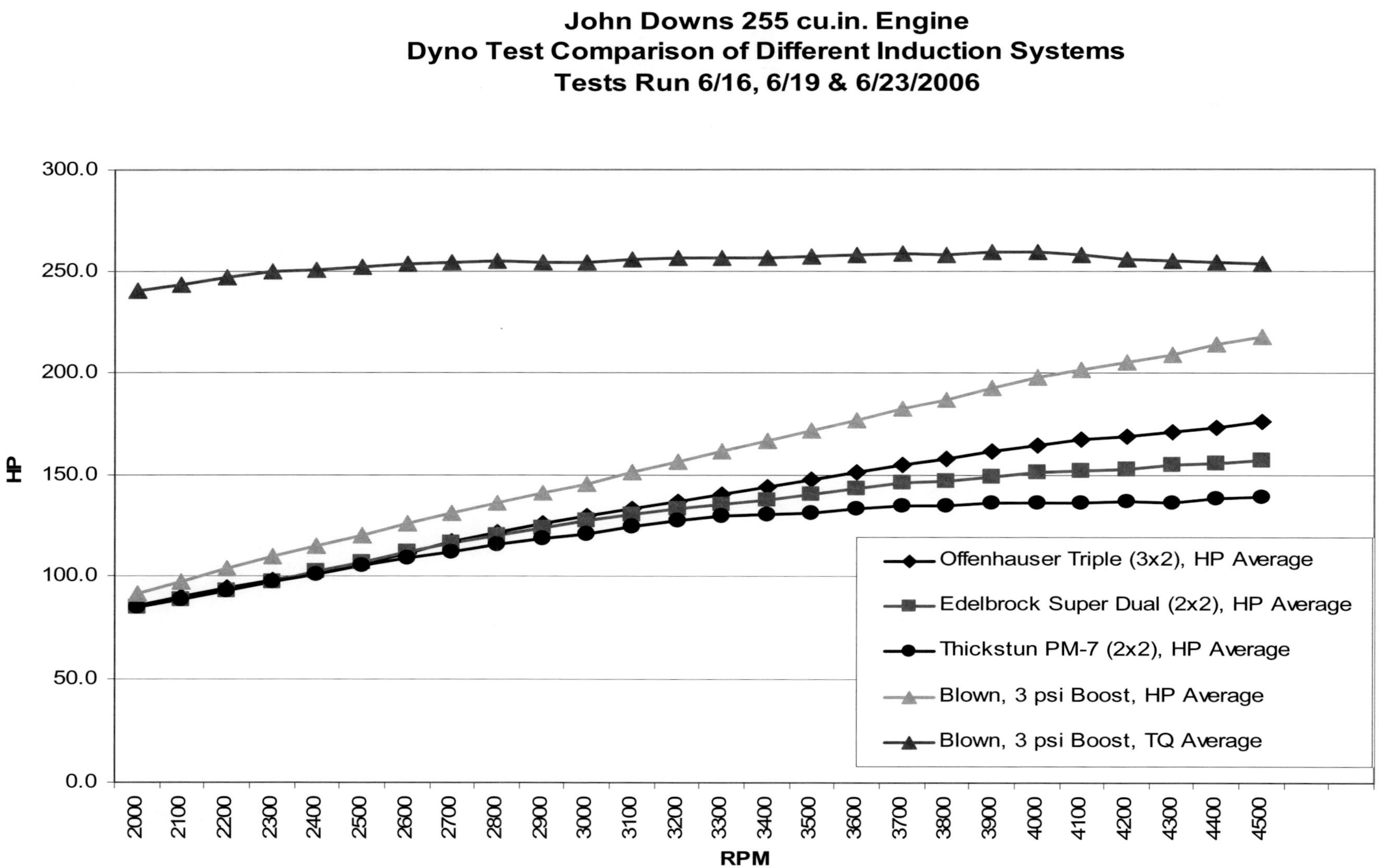

Figure 7-5. Horsepower and Torque vs. RPM for the John Downs 255 cu.in. Flathead Ford. The various induction systems are described in the text. Note the remarkably constant torque line for the supercharged engine.

-8-

Honest Charley-Coker Tire 284 cu.in. Flathead Ford

Figure 8-1. The "Double Exposure Deuce" Ready to Roll. The passenger side has Coker Tire markings and the driver's side has Honest Charley markings accounting for the "double exposure" name. "Honest Mike" Goodman of Honest Charley was the project lead.

SUMMARY

This was the most powerful naturally aspirated street flathead tested to date by Roadrunner Engineering! With an Offenhauser 3x2 manifold, this very stout engine produced peak power readings of 220 horsepower at 5000 rpm.

Supercharged, the engine produced 280 hp and 295 hp at 4 psi and 5 psi boost, respectively. The torque was remarkably high and constant over the entire test rpm range at both boost levels. The boost levels were conservatively limited to allow safe operation at sea level with the 7.9:1 compression ratio and 91-93 octane fuel.

TEST DESCRIPTION AND RESULTS

The engine was built as a joint development project sponsored by the Honest Charley Speed Shop, the Coker Tire Company, Chattanooga, TN and Roadrunner Engineering, Albuquerque, NM. The goal was to produce a strong, reliable street engine using off the shelf parts for the Honest Charley - Coker Tire 1932 Ford roadster.

The engine was subjected to 30+ dyno runs ("pulls") over the 5-day test period. The first series of pulls was performed normally aspirated to break-in the engine, do initial tuning and check for proper function of all components. An Offenhauser ***Triple*** intake manifold with Stromberg ***97*** carburetors prepared by Jere Jobe of Vintage Carburetion Technology Inc. was used for the normally aspirated tests.

Test rpm was conservatively limited to 5000. Peak power for all induction systems occurred at the 5000 maximum test rpm indicating an excellent free breathing engine with potentially more on tap. All testing was performed with an alternator and the water pumps installed using 91-octane premium pump gas. All power values are corrected to standard dyno (sea level) conditions of 60 F and 29.92 in. Hg. This was a very strong 284 cu.in. flathead engine,

Figure 8-2. Ready to Rip! The Honest Charley-Coker Tire supercharged 284 cu.in. flathead shown on the dyno. 295 hp on 91-octane!

both naturally aspirated and supercharged. Measured torque and horsepower curves for this engine are shown in Figure 8-3 at the end of this chapter.

The engine broke in quickly and ran strongly during all tests. However, a failure of the dyno data computer interrupted testing and caused a four-day delay and required a re-run of the normally aspirated testing. The delay allowed a replacement of the one-piece front oil seal on the engine, which evidenced a leak during break-in testing. The seal was replaced with the stock-type rope seal with no further leaks. The stock flathead oil pan and timing gear cover do not properly support a one-piece seal.

The three-carburetor Offy manifold with stock Stromberg ***97*** carburetors operated flawlessly for the normally aspirated testing. No stumbling or hesitation was observed even with the straight (non-progressive) linkage. The carburetors used stock #45 main jets for most tests, which caused the full throttle air-fuel mixture to be ideal (12.5:1) at the 5300 ft test altitude and contributed to the excellent power readings. Peak power normally aspirated was observed with the total timing set at 25 -27 degrees.

For the supercharged testing, peak power was observed at 17-23 degrees total advance with very little difference in output over this range. Running on the low side of this range has the advantage of reduced detonation tendencies and lower octane requirements for the street. Jetting was not varied from stock for the Holley ***Street Avenger*** four barrel since the air-fuel mixture was ideal (11.5:1) for peak-supercharged power at the 5300 ft test altitude. The rich mixture does give an intercooling effect and reduces detonation tendencies. The air-fuel mixture will be close to 12.5:1 (at full throttle) at the normal near sea level operating altitude. Enlarging the secondary jets is recommended for peak power at lower altitude. The Holley ***Street Avenger*** 4-bbl carburetor has vacuum operated secondary throttle plates, which are controlled by an easily replaced spring (2 screws) to control secondary throttle operation to suit driver preference.

Tests were run with 3.75" and 3.48" diameter supercharger pulleys resulting in supercharger overdrive ratios of 50% and 62% respectively with the 5.63" crankshaft pulley. These ratios produced corresponding peak boost readings of 4 psi and 5 psi at full throttle. The engine was shipped with the lower boost pulley installed to assure fuel compatibility during cross-country driving at low altitude.

Following the completion of dyno testing, the engine was installed in the Honest Charley Speed Shop-Coker Tire Co. 1932 Ford roadster, the “Double Exposure Deuce,” and driven from Hamilton, OH to the SEMA show in Las Vegas, NV by “Honest Mike” Goodman and Corky Coker. It performed flawlessly on the 2000+ mile journey and was exhibited on the show floor as driven. The “Driven Dirty Tour” was repeated again in 2007 and the car again made the trip with no problems.

Overall, this engine performed like a champ on the dyno and on the street. As of October 2007, the car and engine have accumulated over 16,000 miles.

The engine build information follows.

ENGINE DESCRIPTON AND SPECIFICATIONS

8BA Short block assembly

a. 3-5/16" bore, Ross forged pistons with stock dome and ductile moly top ring. Piston-cylinder clearance was .006". Top ring gap was .015" and second ring gap was .017".

b. 4-1/8" stroke Scat crank, mild steel center main cap. Main bearing clearance was .002". Crankshaft end play was .004".

c. Eagle H-beam rods. Rod bearing clearance was .0015". Rod side clearance was .014".

d. Speedway 1.57" stainless intake and exhaust valves, block ported and relieved. Valve clearance was 0.012" (intake) and 0.014" (exhaust).

e. Melling ***M15*** high volume oil pump.

f. Isky ***Max1*** cam with single Isky ***185G*** valve springs, shimmed as necessary to get proper installed height.

Heads

Modified Offenhauser ***425*** aluminum heads with a 7.9:1 compression ratio (72 cc head volume) were used. This is optimized for a supercharged flathead designed to operate at sea level with moderate boost and premium pump gas. The head was opened up in the valve pocket area only using a specially ground end mill and then blended by hand as discussed in Chapter 4.

Intake-Naturally Aspirated

Offenhauser ***Triple*** as noted previously. Three stock Stromberg ***97*** carburetors with #45 main jets and #65 power valves.

Intake-Supercharged

A Roadrunner single-plane blower manifold and a Weiand ***142*** blower with 50% and 62% overdrive pulleys were used for the supercharged tests. A 570 cfm Holley ***Street Avenger*** (4-bbl) carburetor with vacuum secondaries and stock jets will be used on the vehicle and was used for most of the testing. Three vacuum secondary springs were furnished with this carburetor. The spring choice for street use will be based on driver preference to tailor low-end torque characteristics, but will not affect peak power readings.

Testing at 62% overdrive was also carried out with a Holley 600 cfm, blower-calibrated carburetor, P/N 0-80592S. No significant difference in power was noted between the two four barrels. Any difference would be expected in drivability which could not be evaluated on the dyno. No air cleaner was used during any of the testing. A large capacity air cleaner is recommended in service.

Exhaust

Free flowing center-dump headers and large (4" inlet/exhaust) truck mufflers were used during test. The engine had no center exhaust baffles.

Ignition

A top mount electronic Mallory distributor (***Unilite***) with mechanical advance and Autolite ***4093*** plugs gapped at .025"-.030" were used for test. The distributor had 16 degrees advance built-in, which was all in by 3600 rpm. Total advance curve was limited to 23 degrees maximum for the supercharged runs. Peak supercharged power occurred at 17-20 degrees total advance. An ignition retard device such as those available from MSD (P/N's 5462 or

6462) with a retard setting of 1 to 1-1/2 degrees per psi of boost is recommended for the street to allow the total timing to be set to 24-26 degrees for part load cruising economy.

Water pumps

Stock ***8BA*** type new pumps with modern seals and bearings were used along with Roadrunner serpentine pulleys and hubs for all tests.

Belt Tensioner

An experimental automatic belt tensioner was used for the supercharged testing. The normal 1-belt Roadrunner supercharger kits use a manual belt tensioning system. The automatic unit maintained the belt tension at 180-185 pounds, which is a little lower than the desired 180-200 pounds specified for the manual unit, but provided better belt wrap for the crank pulley. It was left in place for street evaluation in the project vehicle. No problems to date.

PROJECT PARTICIPANTS

Project lead was Mike Goodman of the Honest Charley Speed Shop. Joe Abbin of Roadrunner Engineering was the project manager. Automotive Machine Service in Albuquerque, NM did the machine work including balancing. Jeff Andreen and Kent Hendershot, Denver, CO did the porting and relieving of the block. The porting on this block was identical to that of the ***Killer*** flathead in Chapter 10. Gary McGlasson of McGlasson Racing Engines, Los Lunas, NM built, tested and tuned the engine. Gary also performed the head modifications.

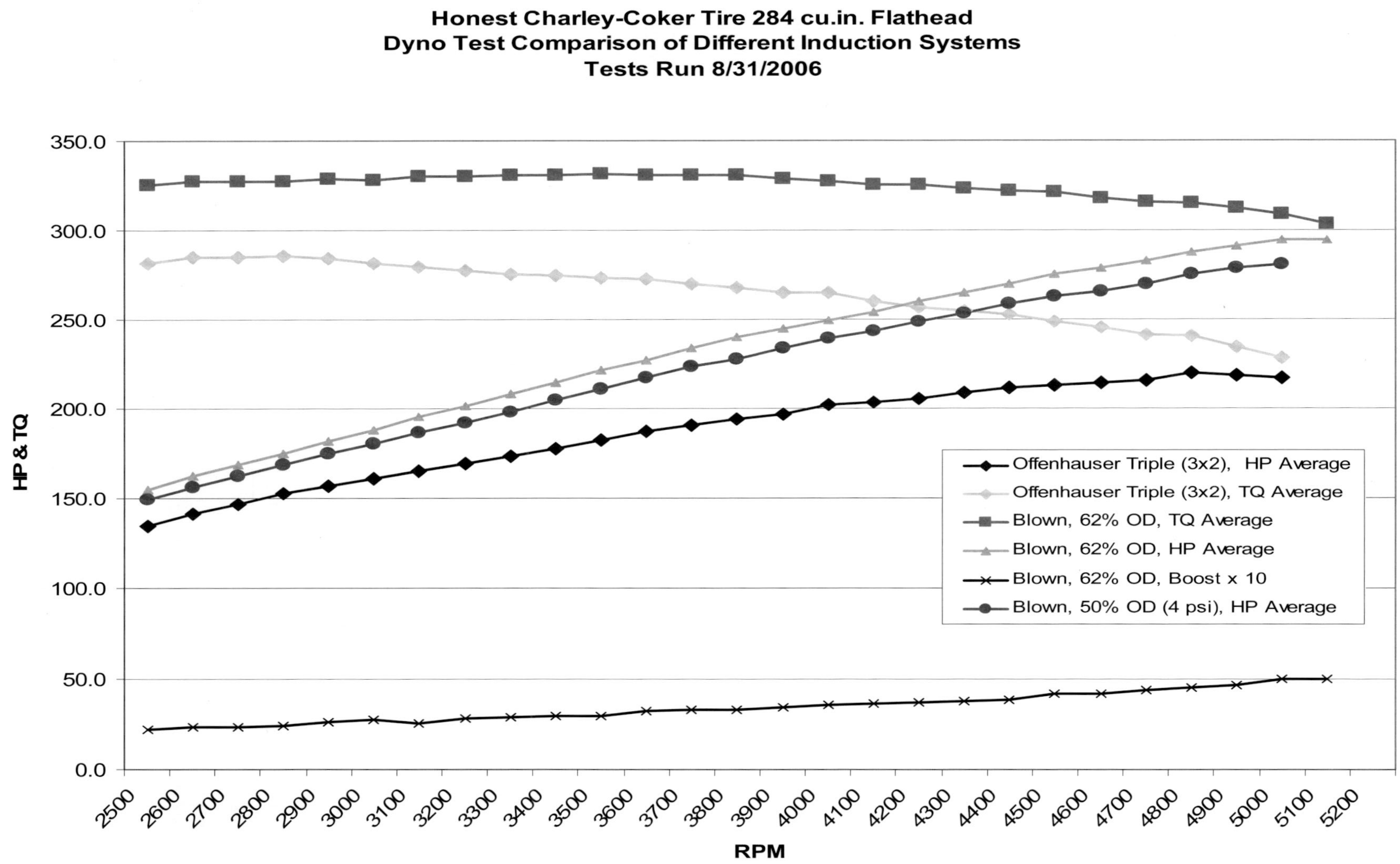

Figure 8-3. Horsepower and Torque vs. RPM for the Honest Charley – Coker Tire 284 cu.in. Flathead Ford. Both normally aspirated and supercharged results are shown. Each data point is an average of three readings.

-9-

Bill Boomer 284 cu.in. Flathead Ford

Figure 9-1. The Boomer 284 cu.in. Blown Flathead on the Dyno. It averaged 305 hp at 5200 rpm and 328 ft-lbs at 3700 rpm with about 5 psi boost maximum!

SUMMARY

The subject engine was built as a street engine development project sponsored by the owner, Bill Boomer of Regina, Saskatchewan, Canada. This engine is similar in many ways to the Honest Charley-Coker Tire engine described in Chapter 8. The Boomer engine demonstrated 217 hp normally aspirated and averaged 305 hp supercharged on the engine dyno. Figure 9-6 compares this engine with the Honest Charley-Coker Tire engine.

TEST DESCRIPTION AND RESULTS

The primary goal of this project was to develop, build and test a 300 hp, reliable flathead street engine at a reasonable cost.

The following paragraphs describe the subject engine and summarize the results of dynamometer testing of the engine. Measured torque and horsepower curves for the very strong 284 cu.in. flathead engine, both naturally aspirated and supercharged, are shown in a plot at the end of this chapter. Test rpm was conservatively limited to 5000 during most of the dyno testing. Peak power for all induction systems occurred at or near the maximum test rpm indicating an excellent free breathing engine with potentially higher capability. All dyno testing was performed with a drive belt, alternator, and water pumps installed and the exhaust running through 4" inlet/outlet truck mufflers. The fuel was 91 octane gasoline for most of the tests. All measured power values are corrected to standard dyno (sea level) conditions of 60 F and 29.92 in.Hg.

The engine broke in quickly and ran strongly during all tests. The planned testing resulting in the engine being subjected to 30+ dyno runs ("pulls") over the test period. The first series of pulls was performed normally aspirated with a stock intake manifold with a single Ford (Holley ***94-*** type) carburetor to break-in the engine, do initial tuning and check for proper function of all components. The engine demonstrated good ring seal and quick break-in during these initial shake down runs on the dyno. For the second series of tests, an Offenhauser ***Triple*** intake manifold and three Stromberg ***97*** carburetors prepared by Jere Jobe of Vintage Carburetion Technology Inc. were used. This was the same setup used during normally aspirated testing of both the Honest Charley-Coker Tire engine (Chapter 8) and "Killer" (Chapter 10). The three-carburetor setup operated flawlessly with no stumbling or hesitation observed even with the straight (non-progressive) linkage. As in previous testing with this manifold and carburetors, stock #45 main jets were used for testing initially. This caused the mixture to be somewhat rich (12.2:1) at the 5300 ft test altitude. The jets were then replaced with #43 units and a series of pulls revealed the air-fuel ratio went up to 12.7:1, but the power was essentially identical. Peak power normally aspirated was observed with total ignition advance set at 24 degrees, although there was little difference down to 20 degrees. The ***Triple*** manifold produced averaged peak power readings of 217 horsepower as shown in the graph. This was a very powerful naturally aspirated street flathead!

Supercharged, the engine produced 280, 290, and 300 hp at 3, 4, and 5 psi boost, respectively. The torque was remarkably high and constant over the entire test rpm range at all boost levels. All boost levels would allow safe operation with 91 octane fuel at the 5300 ft test altitude with the 7.3:1 compression ratio. However, the supercharged tests at the 5 psi boost level were run using 101-octane Trick unleaded gasoline to be conservative while the timing was varied. The full throttle air-fuel ratio averaged approximately 11.2:1 and probably could have been leaner for higher power at high altitude. Jetting was not varied from the stock #54 primary jets and #66 secondary jets for the Holley 570 ***Street Avenger*** four-barrel because the air-fuel mixture was conservative for peak supercharged power at the test altitude, and would be naturally leaner by about one point (but still safe) at sea level. As stated previously, rich mixtures are favored for supercharged engines because they provide an inter-cooling effect and reduce detonation tendencies with little penalty in power.

For the supercharged testing, peak power was observed at 24 degrees total advance with very little difference in output over the range of 18-24 degrees at the boost levels tested. Running on the low end of the range minimizes octane requirements and maximizes detonation resistance, but tends to load up the spark plugs.

Tests were run with a 5.63" crankshaft pulley and three different supercharger pulleys, 3.75", 3.48", and 3.23". These combinations resulted in supercharger overdrive ratios of 50%, 62% and 74%, with the 3, 4, and 5 psi peak boost readings noted above. Each pulley change increased peak boost by about 1 psi and peak power by about 10 horsepower in round numbers. The engine was left with the lower boost pulley installed to give a big margin against detonation with a relatively small penalty in performance.

The engine build information is listed below.

ENGINE DESCRIPTON AND SPECIFICATIONS

8BA Short block assembly

a. The block was thoroughly cleaned and checked for cracks. It was then bored and honed with deck plates (modified stock cast iron heads) and main caps installed and properly torqued to enhance bore roundness and ring seal in service. Finish honing was done with a 400 grit stone. A heavy-duty mild steel center main cap from Doug King was installed, and the main bearing bores were align bored. The block was ported and relieved (0.125" depth) to Roadrunner "max street" definition. The top of the valve guide was shortened to prevent protrusion into the port after porting. Intake ports were port matched to the manifold. All bolt holes were tapped and repaired as needed. Block lifter bores were drilled to aid lifter adjustment. Block deck surface was squared to the crankshaft centerline.

b. Race prepped Scat rods were used. Rod bearing clearance was 0.0021". These rods cleared the block without grinding. Rod side clearance was 0.023".

c. Ross forged, three ring pistons, 3-5/16" bore with stock dome, P/N 99817 were used. The rings were ductile iron, 1.5, 1.5, and 3 mm thick with moly-faced top rings, P/N R3315. The piston-cylinder clearance was 0.006". The top and the second ring gaps were 0.018". The average piston to head clearance in the "squish" area was .090". The bob weight for balancing was 1498 grams.

d. A 4.125" stroke Scat crankshaft with 2.00" rod journals was used. The main bearing clearance was 0.0023" and the crankshaft end play was 0.004".

e. The valves were Manley ***StreetFlo,*** 1.6" diameter stainless intake and exhaust, P/N 10765. The intake and exhaust valve gap was .018".

f. The cam was an Iskenderian ***400Jr*** with single ***185G*** Isky valve springs, shimmed as necessary to get a 2.00" installed height (70 lb. seated spring force). The cam was installed straight up.

g. A Melling ***M19*** stock volume oil was used.

Heads

The heads were modified Mercury "***EAC***" iron heads cut for extra valve clearance for the ***400Jr*** cam with an 81 cc head combustion chamber volume. The resulting compression ratio was 7.3 to 1 for this 284 cu.in. engine. This is optimized for a supercharged flathead designed to operate at sea level with moderate boost and 91 octane pump gas. The head was opened up in the valve pocket area only using a specially ground end mill (1.875" dia. with a 0.125" corner radius) and then hand blended. Best ***Graphtite*** big bore head gaskets were used, P/N 536-G1 and G2. Head retainer nuts were torqued to 55 lb-ft (re-torqued after five hot/cold cycles).

Intake-Naturally Aspirated

An Offenhauser ***Triple*** manifold was used as noted previously. Three stock Stromberg ***97*** carburetors were used, all with #45 main jets for part of the tests and #43 for the rest. #65 power valves were used for all testing with these carburetors.

Intake-Supercharged

A Roadrunner single-plane blower manifold and a Weiand ***142*** blower with 50%, 62% and 74% overdrive pulleys were used for the supercharged tests. A stock 570 cfm Holley ***Street Avenger***, 4-bbl carburetor, (P/N 0-80570) was used for all of the supercharged testing. No air cleaner was used during testing. A large capacity air cleaner will be used in service.

Exhaust

Free flowing center-dump headers and large (4" inlet/exhaust) truck mufflers were used during test. The engine had no center exhaust baffles.

Ignition

A top mount, dual point Mallory distributor with mechanical advance and Autolite ***215*** plugs gapped at .035" were used for test. The distributor had 15 degrees advance built-in, which was fully advanced by 3600 rpm. The total advance curve was limited to 24 degrees maximum for the supercharged runs. Peak supercharged power occurred at 18-24 degrees total advance for all boost levels. An ignition retard device available from MSD (P/N 6462) with a retard setting of 1 to 1-1/2 degrees per psi of boost is recommended for the street to allow the total distributor timing to be set to 24-26 degrees for part load cruising economy.

Water pumps

Stock ***8BA*** type new pumps with modern seals and bearings were used along with Roadrunner serpentine pulleys and hubs for all tests.

Belt Tensioner

A prototype Roadrunner automatic belt tensioning system was used and required no attention during testing. See Figure 9-2.

PROJECT PARTICIPANTS

The author developed the overall engine definition and performed all computer modeling and flow testing. McGlasson Racing Engines, Los Lunas, NM and Automotive Machine Service in Albuquerque, NM did the machine work and balancing. Gary McGlasson did the excellent porting and relieving of the block and assembled the engine. Gary also dyno tested the engine and assisted the author in tuning the engine.

Figure 9-2. A Prototype Roadrunner Automatic Tensioner. This is a billet bracket similar to the cast bracket used on the Honest Charley-Coker Tire engine.

Figure 9-3. The Relieved Area of the Block. The relief is approximately .125" - .188" deep and follows the head gasket boundary. It is not clear in this photo, but there is a chamfer where the relief meets the bore. The valve pockets or bowl areas have been opened up with a 78 degree "bowl hog" to accommodate the 1.6" valves prior to the three-angle valve seat work and hand blending.

Figure 9-4. A 1.785" end mill with a .125" corner radius was used to open up heads over the valves as described in Chapter 4. In this case the cut was approximately .06" deep parallel to the head gasket surface. Additional crescent shaped cuts were made for additional valve clearance. The resulting average head combustion chamber volume was 81 cc's.

Figure 9-5. The Boomer Engine on the Dyno with the Three-two Barrel Carburetor Setup. The engine developed 215 hp average at 5000 rpm and 268 ft-lbs at 3100 rpm!

Test Date	Peak Torque	@ RPM	Average HP 2500-5000	Peak HP @ 5000 RPM	Total Timing	Main Jets	Air/Fuel Average
Naturally Aspirated, Offenhauser Triple with Stromber 97's							
8/29/2008	262	3000	175	213	22	0.045	11.52
8/29/2008	264	3200	175	217	24	0.045	11.48
9/1/2008	264	3000	176	214	22	0.043	
9/1/2008	267	3200	178	216	24	0.043	12.21
9/1/2008	268	3100	177	215	26	0.043	12.11
Weiand 142 Supercharger (3.75 dia. pulley, 50% OD) with Holley 570 Street Avenger							
9/2/2008	310	3700	192	279	18	.054/.066	12.10
9/2/2008	306	3600	192	278	20		
9/2/2008	308	3600	194	281	22		
Weiand 142 Supercharger (3.48 dia. pulley, 62% OD) with Holley 570 Street Avenger							
9/2/2008	316	3800	198	288	18	.054/.066	12.06
9/2/2008	314	3800	199	289	20		
Weiand 142 Supercharger (3.23 dia. pulley, 74% OD) with Holley 570 Street Avenger							
9/2/2008	328	3700	207	299	18	.054/.066	12.11
9/2/2008	328	3700	207	301	20		

Notes:
1) Test results are an average from two or more tests (dyno "pulls").
2) Test results are corrected to 60F, 29.92 in.Hg and dry air.
3) Average peak HP was 310 at 5200 RPM during special testing on 9/16/08.

Table 9-1 Summary of Bill Boomer 284 cu.in. Engine Dyno Test Results. Note peak performance values as the total timing and air-fuel ratio are varied for the normally aspirated tests. For the supercharged tests, only the total timing and supercharger overdrive values were varied.

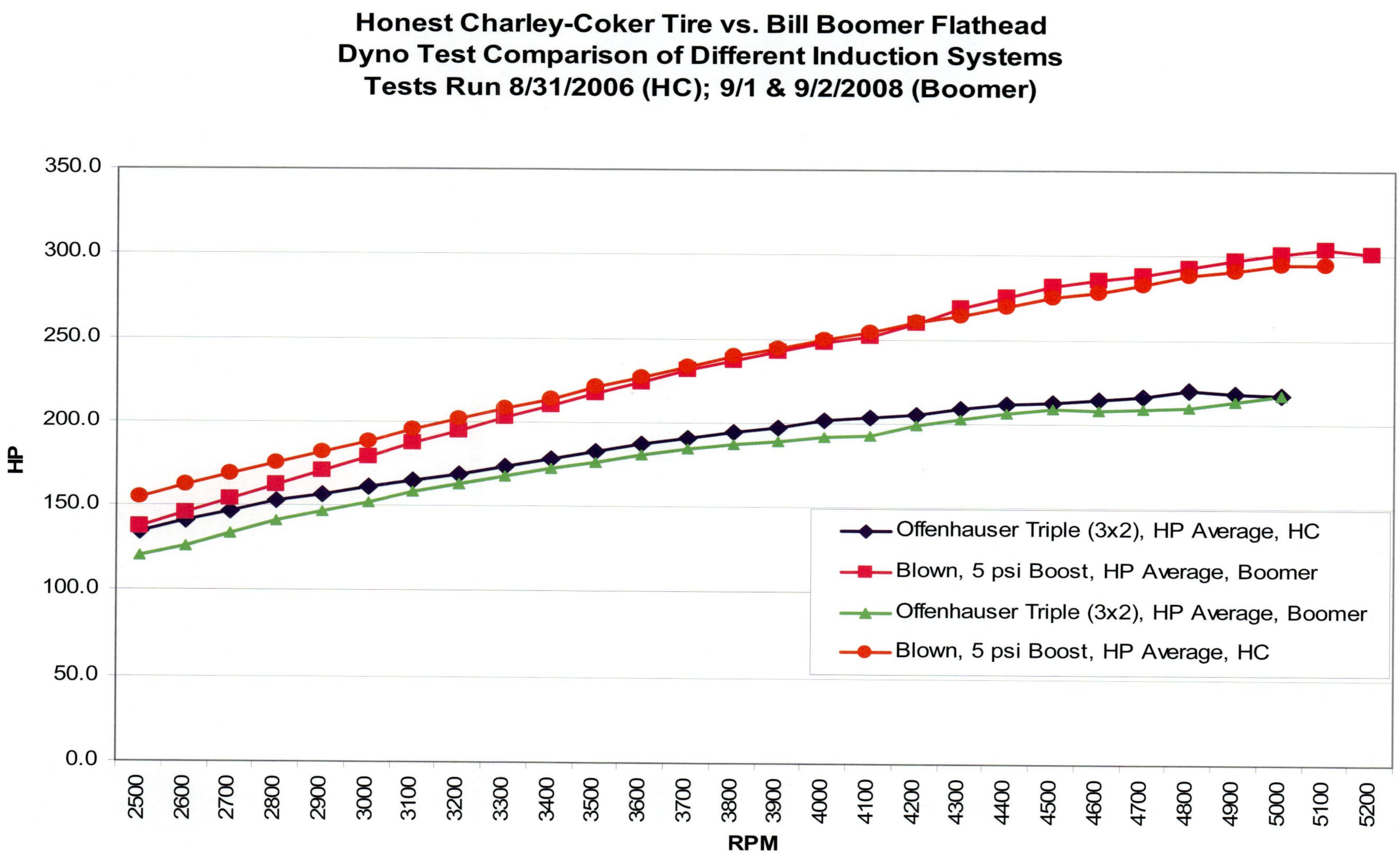

Figure 9-6. Horsepower and Torque vs. RPM Comparison for the Bill Boomer and the Honest Charley-Coker Tire (abbreviated "HC" above) 284 cu.in. Flathead Fords. Each data point is an average of three readings.

-10-

Roadrunner Engineering 276 cu.in. Flathead Ford, aka "Killer"

Figure 10-1. *Killer*, the Roadrunner 276 cu.in. Blown Flathead on the Dyno. *Killer* made 335 hp at 5000 rpm and 365 ft-lbs at 4000 rpm! This may be the most powerful flathead engine ever built for the street! A Weiand *174* supercharger was used.

SUMMARY

The subject engine was built as a development project sponsored and managed by the author, Joe Abbin of Roadrunner Engineering, Albuquerque, NM. This engine is similar in many ways to the first Roadrunner blown flathead built in 1994 and documented in the book, ***Blown Flathead***. That first engine accumulated almost 2000 street miles and over 300 drag strip passes over the ten-year period that the engine was installed in the ***Motorhead Mart Special***. During that time, a wealth of data and experience was accumulated and incorporated into the next generation engine described in this chapter. The new engine demonstrated 217 hp normally aspirated, and averaged 335 hp supercharged on the Roadrunner/McGlasson engine dyno. This outstanding performance was verified on the drag strip after the engine was installed in the ***Motorhead Mart Special***. The new combo demonstrated best performances of 107.6 mph for top speed and 12.41 seconds for elapsed time. These drag strip performances bettered the car's previous bests at the same location with the previous engine by more than 7 mph in top speed with a full second lower elapsed time. The 0-60 time went from about 4.5 seconds to 3.5 seconds!

TEST DESCRIPTION AND RESULTS

The primary goal of this project was to develop, build and test a strong and reliable flathead street engine at a reasonable cost. A second goal was to characterize and optimize each stage of the engine buildup and further refine the ***Engine Analyzer Pro*** computer models. Valid computer models facilitate the design, evaluation and optimization of future engine designs with minimum hardware and test investment.

The following paragraphs describe the subject engine and summarize the results of dynamometer testing and analyses of the subject engine. Additional pictures and detailed individual tabular data sheets from the computer modeling and the dyno testing are shown later in the chapter. Measured torque and horsepower curves for the very strong 276 cu.in. flathead engine, both naturally aspirated and supercharged, are shown in a plot at the end of this chapter. Test rpm was conservatively limited to 5200 during dyno testing. Peak power for all induction systems occurred at or near the 5200 maximum test rpm indicating an excellent free breathing engine with potentially higher capability. All dyno testing was performed with a drive belt, alternator and water pumps installed and the exhaust running through 4" inlet/outlet truck mufflers. The fuel was gasoline of various octane ratings as noted. All measured power values are corrected to standard dyno (sea level) conditions of 60 deg F and 29.92 in. Hg.

The engine broke in quickly and ran strongly during all tests. However, numerous glitches associated with the dyno and data acquisition setup interrupted and extended testing over a six-week period and required numerous re-runs of the planned testing resulting in the engine being subjected to 80+ dyno runs ("pulls") over the test period. These delays did allow time for detailed analyses of the data in between tests, which was a good thing. The computer models of the engine were refined as testing progressed and will allow better performance predictions and analyses of this engine and future engines.

The first series of pulls was performed

Figure 10-2. *Killer* engine on the dyno with dual-two barrel carburetor setup. 199 hp!

normally aspirated to break-in the engine, do initial tuning and check for proper function of all components. The engine demonstrated good ring seal and quick break-in after a thirty minute shake down run on the dyno and approximately a half dozen pulls. An Edelbrock ***Super Dual*** and an Offenhauser ***Triple*** intake manifold with Stromberg ***97*** carburetors prepared by Jere Jobe of Vintage Carburetion Technology Inc. were used for the normally aspirated tests. The two and three-carburetor setups operated flawlessly with no stumbling or hesitation observed even with the straight (non-progressive) linkage. The carburetors used stock #45 main jets for most tests, which caused the full throttle mixture to be ideal (12.5:1) at the 5300 ft test altitude and contributed to the excellent power readings. Shamrock 91-octane premium unleaded gasoline was used for these tests. Peak power normally aspirated was observed with total ignition advance set at 23-27 degrees although there was little difference down to 21 degrees. Conversely, advancing the timing to 32 degrees caused a huge power loss (dropped 75 hp!). The dual manifold produced nearly 200 average peak hp and the triple manifold produced average peak power readings of 217 horsepower as shown in Figure 10-14. This was a very powerful naturally aspirated street flathead!

Supercharged, the engine produced 315 hp and 335 hp at 6.3 psi and 8 psi boost, respectively. The torque was remarkably high and constant over the entire test rpm range at both boost levels. The boost levels were limited to allow safe operation at the 5300 ft test altitude with the 8.2:1 compression ratio. Supercharged tests were run using 101-octane Trick unleaded gasoline and 110-octane race gas to be conservative while the timing was varied. The full throttle air-fuel ratio averaged approximately 11.2:1 and probably could have been somewhat leaner for higher power at high altitude based on subsequent road testing. Jetting was not varied from the #70 primary jets and #72 secondary jets for the Holley four-barrel because the air-fuel mixture was conservative for peak supercharged power at the test altitude. The air-fuel ratio would be leaner by about one point (but still safe) at sea level. As noted earlier, rich mixtures are favored for supercharged engines because they provide an inter-cooling effect and reduce detonation tendencies with little penalty in power.

For the supercharged testing, peak power was observed at 16-23 degrees total advance with very little difference in output over this range at either boost level tested. As shown in Tables 10-1 and 10-2, **peak performance was virtually identical at 16 degrees and 20 degrees total advance**. Running on the low end of the range minimizes octane requirements and maximizes detonation resistance, but tends to load up the spark plugs.

Tests were run with a 5.625" crankshaft pulley and two different supercharger pulleys, 4.10" and 3.75". These combinations resulted in supercharger overdrive ratios of 37% and 50%, and peak boost readings of 6.3 psi and 8 psi respectively, as stated previously. The higher boost increased peak power by about 20 hp, averaging over 335 hp at 5000 rpm and 365 lb-ft at 4000 rpm.

The engine was left with the higher boost pulley installed to compete at the Flathead and Inline Nationals at Denver, which is about the same altitude as Albuquerque (5800 ft. vs. 5300 ft.).

The engine was then installed in the author's 1934 Ford sedan, the ***Motorhead Mart Special***, and run on the street and the drag strip at high (5800 ft.) altitude. During competition at Denver in June 2006, the engine pulled the 2540 lb. ***Motorhead Mart Special*** to a quickest ET of 13.28 seconds in the quarter mile and a fastest speed of 100.3 mph. Subsequent runs with the lower boost pulley in October 2006 in Bakersfield, CA (altitude 380 ft.) at the NHRA Hot Rod Reunion netted a quickest ET of 12.41 seconds and a fastest speed of 107.6 mph See Table 2-1 in Chapter 2. These times indicate that the dyno results were no fluke and that the engine is "the real deal." The engine build information follows.

Figure 10-3. Killer on the Dyno with Three-two barrel Carburetion. 217 hp!

ENGINE DESCRIPTION AND SPECIFICATIONS

8BA Short Block Assembly

a. The block was thoroughly cleaned and checked for cracks. It was then bored and honed with deck plates (modified stock cast iron heads) and main caps installed with proper torque to enhance bore roundness and ring seal in service. A heavy-duty mild steel center main cap from Doug King was installed. The block was ported and relieved (0.125" depth) to Roadrunner "max street" definition.

b. Race-prepped stock Ford ***8BA*** rods were used. Rod bearing clearance was 0.0015-0.002".

c. Ross forged, three ring pistons, 3-5/16" bore with stock dome, P/N 99817 were used. The rings were ductile iron, 1.5, 1.5, and 3 mm thick with moly-faced top rings, P/N R3315. The piston-cylinder clearance was 0.005". The top ring gap was 0.015" and the second ring gap was .017". The average piston to head clearance in the "squish" area was .090". The bob weight for balancing was 1502 grams.

d. A 4" stroke, stock Mercury crank with .010" undersize rods and mains was used. Crankshaft end clearance was .006" and main bearing clearance was 0.001-0.0015".

e. The valves were Manley ***StreetFlo,*** 1.6" diameter stainless intake and exhaust, P/N 10765. The intake valve gap was 0.015-0.016" and the exhaust valve gap was 0.016-0.017".

f. The cam was an Iskenderian ***400Jr*** with single Isky ***185G*** valve springs, shimmed as necessary to get a 1.972" installed height (92 lb. seated spring force). The cam was installed .25 degree advanced.

g. A Melling ***M15*** high volume oil pump moves the oil.

Heads

Modified aluminum Offenhauser ***425*** heads with a 67 cc head combustion chamber volume were used. The resulting compression ratio was 8.2:1 for this 276 cu.in. engine. This is optimized for a supercharged flathead designed to operate at sea level with moderate boost and race gas. The heads were opened up in the valve pocket area only using a specially ground end mill (1.875" dia. with a 0.125" corner radius) and then hand blended. Best ***Graphtite*** big bore head gaskets were used, P/N 536-G1 and G2. Head nuts were torqued to 55 lb-ft (re-torqued after five hot/cold cycles).

Intake-Naturally Aspirated

Edelbrock ***Super Dual*** and Offenhauser ***Triple*** manifolds were used as noted previously. Two or three stock Stromberg ***97*** carburetors were installed as required, all with #45 main jets and #65 power valves.

Intake-Supercharged

A Roadrunner single-plane blower manifold and a Weiand ***174*** blower with 37% and 50% overdrive pulleys were used for the supercharged tests. A modified 600 cfm Holley double-pumper, 4-bbl carburetor (P/N 4776) was used for all of the supercharged testing. No air cleaner was used during testing. A large capacity air cleaner is used in service.

Exhaust

Free flowing center-dump headers and large (4" inlet/exhaust) truck mufflers were used during test. The engine had center exhaust baffles.

Ignition

A top mount, dual point Mallory distributor with mechanical advance and Autolite ***4092*** plugs gapped at 0.025"-.030" were used for test. The distributor had 15 degrees advance built-in, which was fully advanced by 3600 rpm. The total advance curve was limited to 23 degrees maximum for the supercharged runs. Peak supercharged power occurred at 16-21 degrees total advance. An ignition retard device available from MSD (P/N 6462) with a retard setting of 1-2 degrees per psi of boost is used for the street to allow the total distributor timing to be set to 24-26 degrees for part load cruising economy.

Water pumps

New ***8BA*** type pumps with modern seals and bearings were used with Roadrunner serpentine pulleys and hubs for all tests.

Belt Tensioner

The normal 1-belt Roadrunner manual belt tensioning system was used and adjusted as necessary to maintain the desired 190-200 pounds tension. Two Roadrunner "wrapper" idlers were required to improve drive belt wrap and prevent slippage at 9 psi boost.

PROJECT PARTICIPANTS

The author developed the overall engine definition, performed all computer modeling and flow testing, and also performed the rod and head modifications. Automotive Machine Service in Albuquerque, NM and Ridge Reamer Machine in Arvada, CO did the machine work and balancing. Jeff Andreen and Kent Hendershot, Denver, CO did the superb porting and relieving of the block and assembled the engine. Gary McGlasson of McGlasson Racing Engines, Los Lunas, NM dyno tested and assisted the author in tuning the subject engine.

Figure 10-4. The block with Doug King's heavy-duty center main cap is shown. Here the main cap inside diameter is being measured. *Jeff Andreen photo*

Figure 10-5. This picture shows the relieved area of the block. The relief is approximately .125" deep and follows the head gasket boundary. It is not clear in this photo, but there is a chamfer where the relief meets the bore. The valve pockets or bowl areas have been opened up with a 78 degree "bowl hog" to accommodate the 1.6" valves prior to the three angle valve seat work and hand blending.

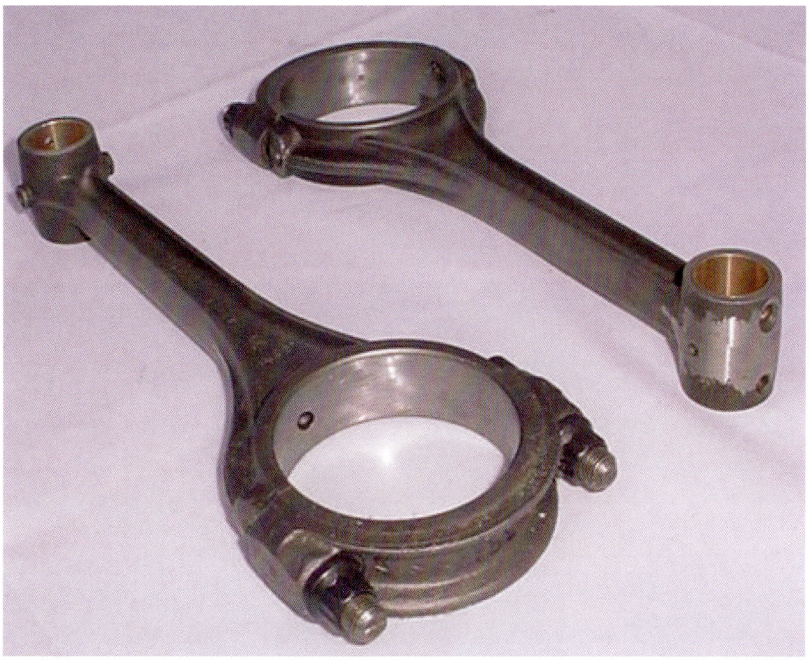

Figure 10-6. The *8BA* rods were modified by smoothing the sides of the I-beams and shot-peening. Note the ARP nuts.

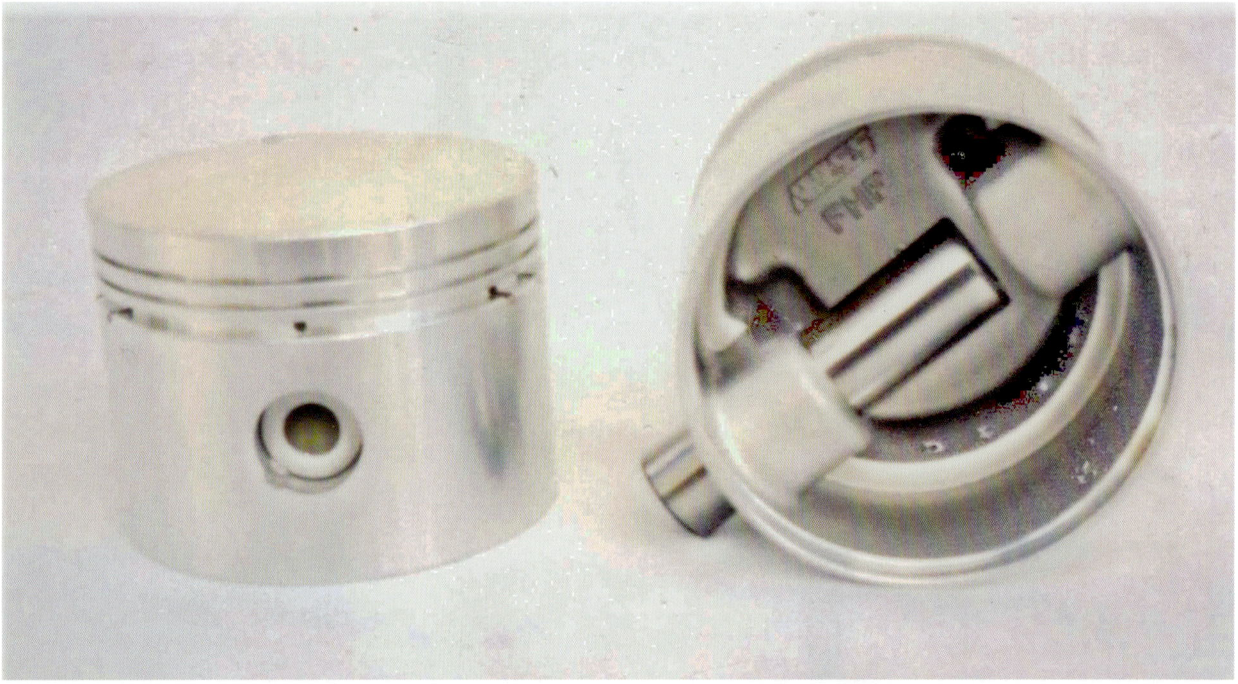

Figure 10-7. Ross Forged Pistons. These were used with a 1.5 mm top ring, a 1.5 mm second ring, and a 3 mm oil ring package.

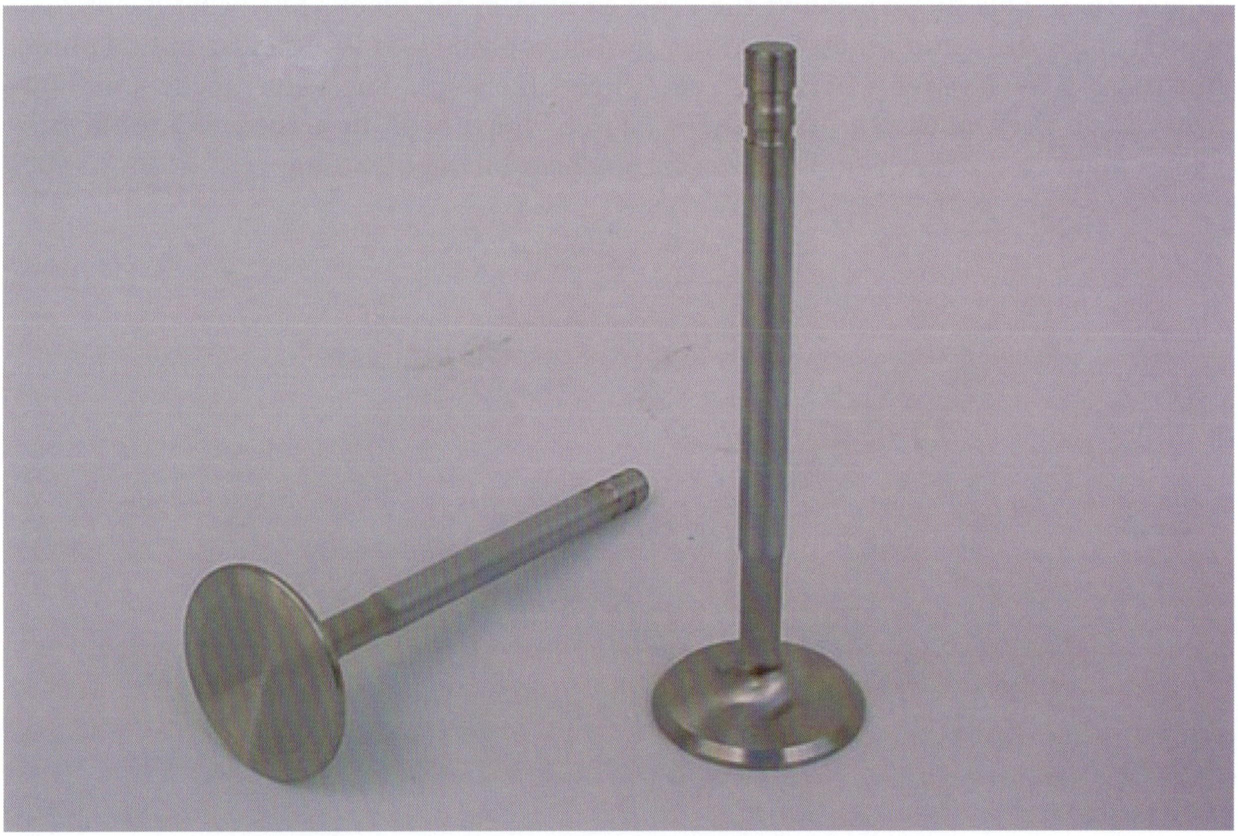

Figure 10-8. Manley *StreetFlo* Valves. These are high performance replacements for stock Chevrolet exhaust valves. 1.6" valves were used for both intake and exhaust.

Figure 10-9. A 1.785" end mill with a .125" corner radius was used to open up the heads over the valves as described in Chapter 4. In this case the cut was approximately .06" deep parallel to the head gasket surface.

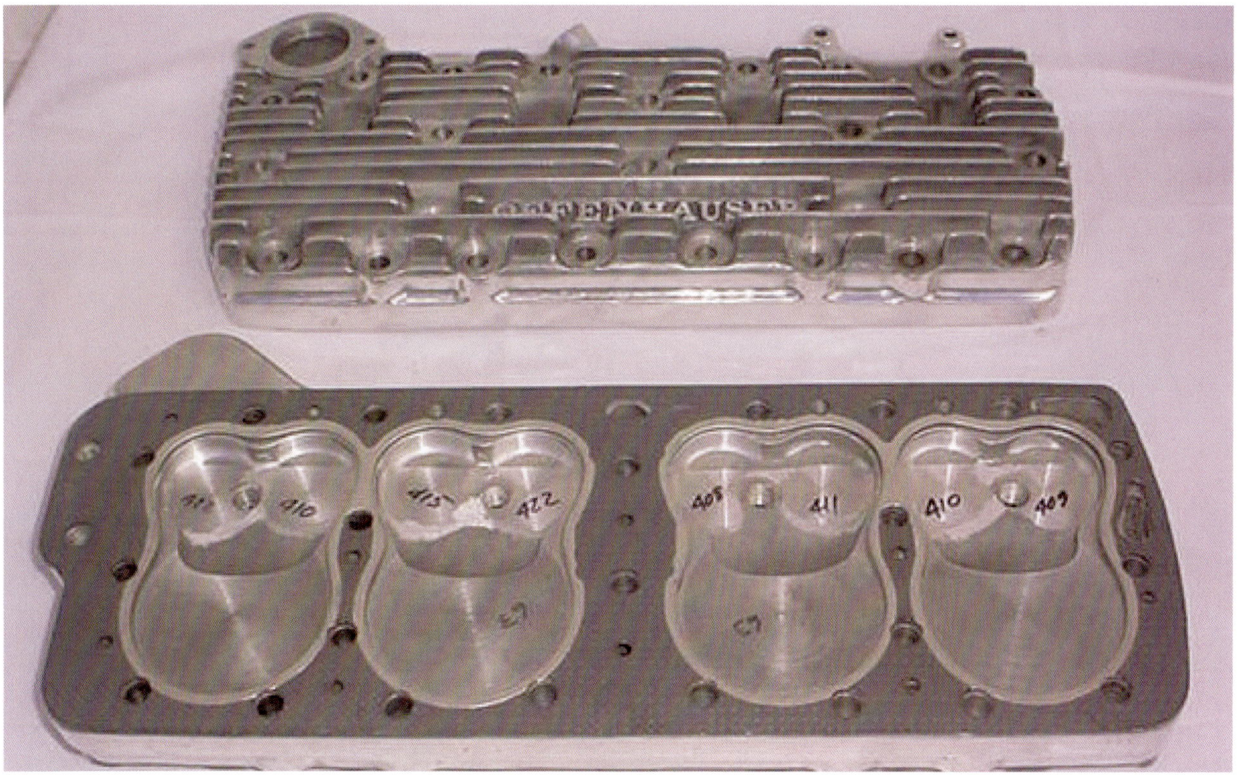

Figure 10-10. After the heads were opened up in the valve area with the end-mill, the milled surfaces were hand blended with the cast surfaces. Note the mill was used to extend the valve pockets as close as practical to the gasket boundaries to minimize shrouding of the valves. The chambers were then polished. The resulting head combustion chamber volume was 67 cc's.

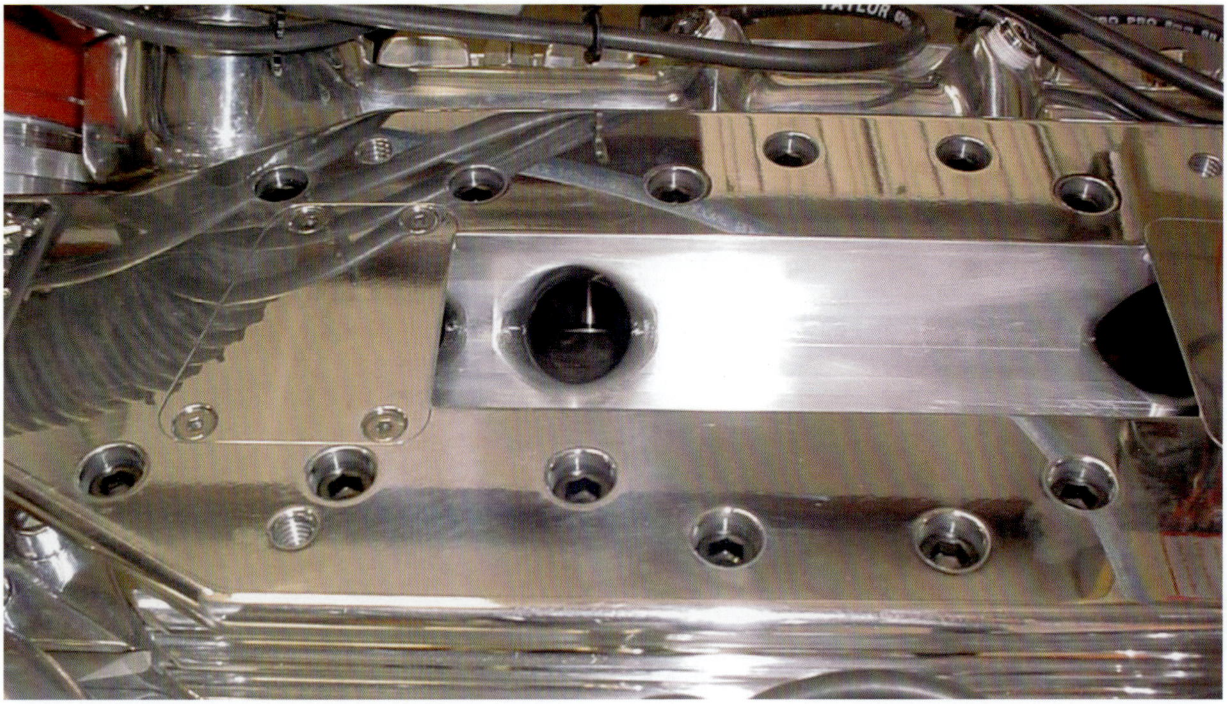

Figure 10-11. The Intake Port in the Roadrunner Machined Billet Intake Manifold.

Figure 10-12. Flow Test Plumbing for the Flathead. The head is in place but not shown. The valves are raised and lowered by the socket head bolts shown directly under the flow duct that pipes air into or out of the cylinder during test. Flow in the duct is measured by an Audie Technology *Flow Quik (*not shown) to produce data like that in Figure 10-13.

Engine Analyzer Pro v3.3 (c) Performance Trends 2004 Intake Flow Table

Intake Flow for 28 Inches Water Pressure 1 Valve, 1.6" dia

Valve Lift, in	Valve L/D	Flow, CFM	Flow Coef
.05	.031	30	.780
.1	.063	54.5	.709
.15	.094	78.4	.680
.2	.125	95.6	.622
.25	.156	108	.562
.3	.188	117	.507
.35	.219	122	.453
.4	.250	123	.400

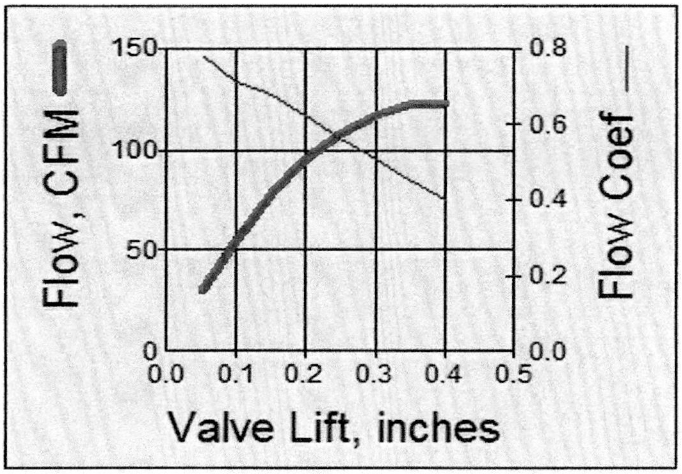

OK | Help | Print Screen | Clear CFM Only | Print Int+Exh | Clear All Data | Print Setup | Factor Up | Factor Down

Tip: It is best to have one point at a Valve L/D of close to .25. Use radiused (or clay) entrance when flowing intake.

14-03-08 02:04 pm Head File:OFFY425A Killer low test results for modified Offenhauser

Engine Analyzer Pro v3.3 (C) Performance Trends 2004 Exhaust Flow Table

Exhaust Flow for 28 Inches Water Pressure 1 Valve, 1.6" dia

Valve Lift, in	Valve L/D	Flow, CFM	Flow Coef
.05	.031	24.2	.629
.1	.063	49.5	.644
.15	.094	67.6	.586
.2	.125	75.6	.492
.25	.156	83.9	.436
.3	.188	88.6	.384
.35	.219	92.1	.342
.4	.250	95.2	.310

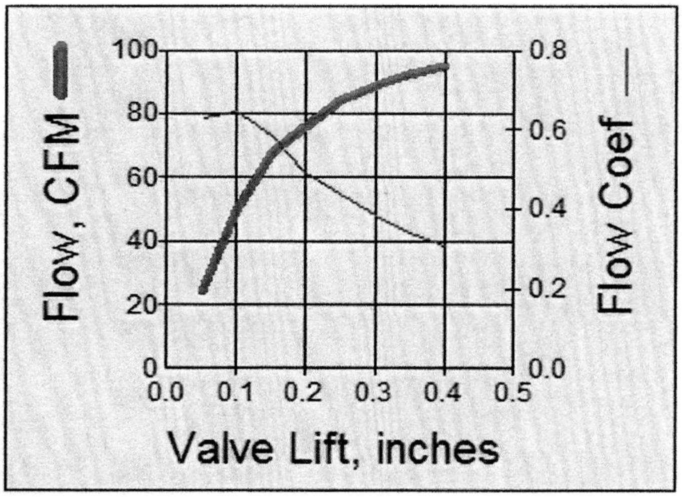

OK | Help | Print Screen | Clear CFM Only | Print Int+Exh | Clear All Data | Print Setup | Factor Up | Factor Down

Tip: It is best to have one point at a Valve L/D of close to .25. Use port extension pipe when flowing exhaust. See page 103.

Figure 10-13. Flow Test Results for the Roadrunner *Killer* Engine. The engine was prepared as noted in the text with a "maximum street port and relief", 1.6" valves for intake and exhaust, and modified Offenhauser *425* heads with 67 cc head volume. The flow test results are input into *Engine Analyzer Pro* as shown above and are used along with other inputs by the program to predict engine performance.

Table 10-1. Roadrunner "Killer" Engine Testing on 2-3-2006 @ 18-51-14

Total Ignition Advance, 16 degrees (Compare with Table 10-2, no notable change in Torque & HP)

Air Temp, 53 degrees; Humidity, 32%; Barometric Pressure, 24.5 in.Hg
Correction Method: Standard (60F, 29.92 in.Hq, dry air)
Weiand 174 supercharger, 50% OD; Holley 4776 carburetor, #70 pri, #72 sec jets, 85 power valve.

RPM	Observed TQ (ft-lb)	Corrected TQ (ft-lb)	Corrected HP (Hp)	BSFC (lb/Hp-hr)	A/F Ratio	Boost (psi)	Center EGT (Degree F)	End EGT (Degree F)
2500	258	326	155	0.55	12.2	4.1	858	738
2600	260	330	163	0.54	12.1	4.2	877	757
2700	262	332	171	0.52	11.8	4.5	880	760
2800	265	337	180	0.51	11.7	4.5	883	768
2900	266	338	187	0.50	11.6	4.8	891	775
3000	267	340	193	0.52	11.5	4.8	901	784
3100	270	344	203	0.54	11.5	4.8	912	794
3200	272	348	212	0.54	11.4	5.0	921	800
3300	274	350	220	0.53	11.4	5.5	931	807
3400	276	353	229	0.53	11.5	5.6	943	817
3500	277	356	237	0.55	11.4	5.5	958	828
3600	278	357	245	0.56	11.4	5.7	977	845
3700	278	357	252	0.56	11.3	6.0	991	855
3800	278	359	260	0.55	11.3	5.8	1002	866
3900	278	359	266	0.54	11.3	6.1	1020	878
4000	275	357	272	0.56	11.3	6.4	1038	882
4100	280	364	284	0.56	11.3	6.5	1057	896
4200	282	**367**	293	0.55	11.3	6.5	1075	914
4300	280	364	298	0.54	11.3	6.9	1085	926
4400	278	363	304	0.54	11.3	7.0	1098	939
4500	276	360	309	0.55	11.3	6.9	1109	948
4600	270	354	309	0.58	11.2	6.8	1125	963
4700	267	351	314	0.58	11.1	7.2	1135	972
4800	267	352	322	0.59	11.2	7.4	1147	980
4900	266	351	328	0.58	11.3	7.5	1164	994
5000	264	350	332	0.58	11.4	7.8	1182	1010
5100	261	346	**336**	0.58	11.4	7.7	1194	1023
5200	253	338	334	0.60	11.4	7.9	1210	1036

Table 10-2. Roadrunner "Killer" Engine Testing on 2-3-2006 @ 18-16-17

Total Ignition Advance, 20 degrees (Compare with Table 10-1, no notable change in Torque & HP)

Air Temperature, 51 degrees; Humidity, 28%; Barometric Pressure, 24.6 in.Hg
Correction Method: Standard (60F, 29.92 in.Hq, dry air)
Weiand 174 supercharger, 50% OD; Holley 4776 carburetor, #70 pri, #72 sec jets, 85 power valve.

RPM	Observed TQ (ft-lb)	Corrected TQ (ft-lb)	Corrected HP (Hp)	BSFC (lb/Hp-hr)	A/F Ratio	Boost (psi)	Center EGT (Degree F)	End EGT (Degree F)
2500	262	326	156	0.54	12.0	4.0	865	755
2600	264	328	162	0.56	11.9	4.3	876	766
2700	266	331	170	0.54	11.7	4.5	880	771
2800	267	333	178	0.53	11.5	4.9	881	779
2900	267	333	184	0.54	11.5	4.8	890	785
3000	267	334	191	0.55	11.5	4.8	900	796
3100	271	339	200	0.55	11.4	5.2	907	804
3200	273	342	209	0.55	11.4	5.4	918	812
3300	274	344	216	0.55	11.4	5.3	928	820
3400	274	345	223	0.56	11.4	5.7	941	828
3500	276	348	232	0.56	11.4	5.8	956	837
3600	279	352	241	0.56	11.2	5.8	971	851
3700	281	355	250	0.54	11.2	6.1	984	863
3800	283	358	259	0.52	11.1	6.0	995	871
3900	286	361	268	0.53	11.2	6.2	1012	880
4000	289	**366**	279	0.52	11.3	6.5	1028	888
4100	286	362	283	0.54	11.2	6.7	1043	899
4200	286	364	291	0.55	11.2	6.6	1057	911
4300	285	363	298	0.54	11.3	6.6	1068	920
4400	283	361	302	0.54	11.3	6.8	1081	935
4500	280	358	307	0.55	11.3	7.0	1093	950
4600	274	351	306	0.58	11.2	7.0	1111	965
4700	271	349	312	0.59	11.1	7.1	1120	977
4800	269	347	317	0.59	11.2	7.4	1131	983
4900	271	350	326	0.57	11.3	7.8	1146	996
5000	269	348	331	0.57	11.3	7.8	1158	1008
5100	265	344	334	0.57	11.2	7.9	1168	1018
5200	261	339	**335**	0.59	11.3	8.1	1179	1026

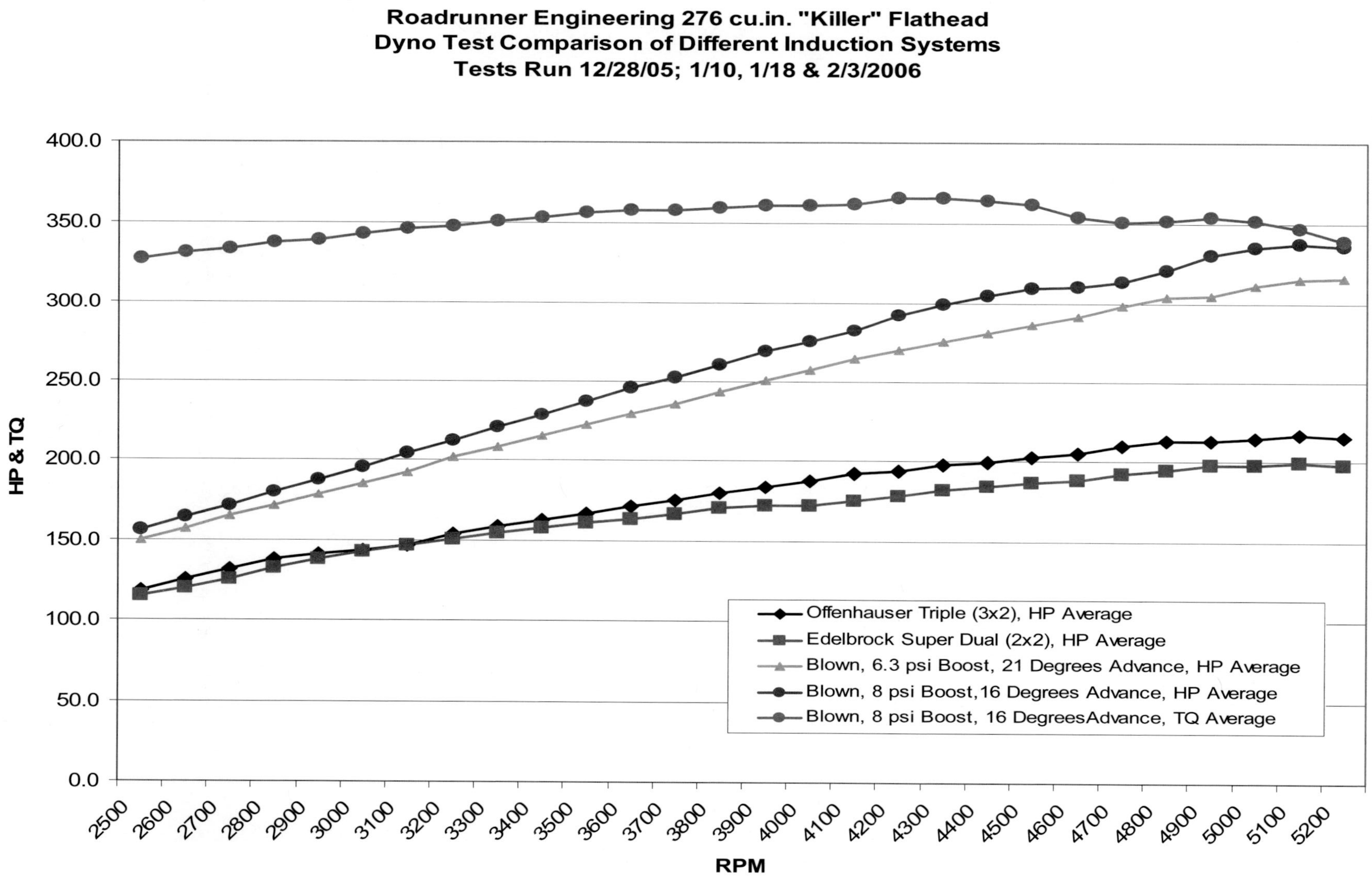

Figure 10-14. Horsepower and Torque vs. RPM for *Killer*, the Roadrunner Engineering 276 cu.in. Flathead Ford. Both normally aspirated and supercharged results are shown. Each data point is an average of three readings.

Appendix A

Directory for Information, Parts, & Services

The suppliers listed below are just a sampling of those who provide parts and services for Roadrunner Engineering and/or the flathead Ford. Many are referenced in the text.

Audie Technology 23 North Trooper Rd. Norristown, PA 19403 Phone: (610) 630-5895 audie_technology@compuserve.com	flow testers and other test equipment
Automotive Machine Service 1645 Second Street NW Albuquerque, NM 87102 Phone: (505) 242-3548	engine machining and rebuilding
Bob Drake Reproductions 1819 NW Washington Blvd. Grants Pass, OR 97526 Phone: (800) 221-3673	body parts, trim and rubber
Car Products PO Box 1 Paragona, UT 84760 Phone: (435) 477-8213	speed equipment
C.W. Moss 402 W. Chapman Ave. Orange, CA 92866 Phone: (800) 322-1932 Fax: (714) 639-0329	engine, chassis and body parts

Cornhusker Rod & Custom
R.R. 1, Box 47
Alexandria, NE
Phone: (402) 749-1932

engine parts, 409 H_20 pump conversions, T-5 transmission adapters, street rod fabrication

Dennis Carpenter Reproductions
P.O. Box 26398
Charlotte, NC 28221-6398
Phone: (800) 476-9653

engine, chassis and body parts, rebuilt and new engines

Early Ford Store
108 West Bonita Ave.
San Dimas, CA 91773
Phone: (909) 305-1955
www.earlyfordstore.com

engine, chassis, and body parts

Egge Machine
11707 Slauson Ave.
Santa Fe Springs, CA
Phone: 866-586-EGGE
www.egge.com

engine parts and speed equipment

Flathead Jack
1561 Third Ave.
Walnut Creek, CA 94596
Phone: (888) 993-2233
Fax: (925) 932-1851
www.flatheadjack.com

engine parts and speed equipment

Flat-O-Products, Inc.
2195 Commercial SE
Salem, OR 97303
Phone: (503) 364-6194

transmission adapters, converters

H & H Flatheads
4451 Ramsdell Ave
La Crescenta, CA 91214
Phone: (818) 248-2371
www.flatheads-forever.com

Navarro racing equipment, S.Co.T. supercharger kits, engine parts and service

Holley Performance Products 1802 Russellville Road Bowling Green, KY 42101 Phone: 270-782-2900 www.weiand.com	engine parts Holley, Weiand, NOS, Hooker, etc. brands
Honest Charley Speed Shop 1309 Chestnut Street Chattanooga, TN 37402 Phone: (888) 795-7077 www.honestcharley.com	speed equipment, body parts clothing, collectibles and memorabilia
Iskenderian Racing Cams 16020 S. Broadway Gardena, CA 90247 Phone: (323) 770-0930 www.iskycams.com	camshafts and valve train components
Joblot Automotive 98-11 211th Street Queens Village, NY 11429 Phone: (800) 221-0172	engine and chassis parts
Land & Sea 25 Henniker St. Concord, NH 03301 Phone: (603) 226 dyno www.land-and-sea.com	engine and chassis dynos
Mac's Antique Auto and Street Rod Parts P.O. Box 1217 Lockport, NY 14095 Phone: (800) 388-7637	engine, chassis and body parts
Mike's Sidevalve Supply 12 Nottage Terrace Medindie Gardens, S.A. 5081 Australia www.flatattackracing.com	books, parts, and speed equipment

McGlasson Racing Engines PO Box 2009 Los Lunas, NM 87031 Phone: (505) 865-0059	engine rebuilding
MBI Publishing Co. 729 Prospect Ave., PO Box 1 Osceola, WI 54020-0001 Phone: (800) 458-0454	books on flatheads, including service manuals
Motorhead Mart PO Box 53296 Albuquerque, NM 87153 Phone/Fax: (505) 268-6768	books on flatheads
Obsolete Ford Parts 8701 South I-35 Oklahoma City, OK 73149 Phone: (800) 654-3247	engine, chassis, and body parts
Patricks P.O. Box 10648 Casa Grande, AZ 85230 Phone: (520) 836-1117 www.patricksantiquecars.com	engine parts and speed equipment
Performance Trends PO Box 573 Dearborn Heights, MI 48127 Phone: (248) 473-9230 www.performancetrends.com	software including ***Engine Analyzer Pro*** and ***Drag Racing Analyzer Pro***
Red's Headers 31410 Reserve Dr. Unit 4 Thousand Palms, CA 92276 Phone: (760) 343-2590 www.reds-headers.com	engine parts and speed equipment

Roadrunner Engineering PO Box 53296 Albuquerque, NM 87153 Phone/Fax: (505) 268-6768 www.roadrunnerengineering.com	engine analysis and supercharger kits
Scat Enterprises, Inc. 1400 Kingsdale Ave. Redondo Beach, CA 90278 Phone: 310-370-5501 www.crankshafts.com	crankshafts, rods, and stroker kits
Speedway Motors P.O. Box 81906 Lincoln, NE 68501 Phone: (800) 979-0122 Fax: (800) 736-3733 www.speedwaymotors.com	major source of parts and speed equipment for street rods and race cars
Vintage Carburetion Technology Inc. 145 Looking East Dr. Somers, MT 59932 Phone: (406) 857-2137 vcti@msn.com	carburetor and distributor parts & service
Wilcap Co. PO Box 763 Pismo Beach, CA 93448 Phone: (805) 481-7639 www.wilcap.com	Sharp speed equipment, heavy-duty main caps, adapters

These pictures depict a small part of Bill Smith's 200+ flathead Ford intake manifolds displayed at his *Museum of American Speed*, located adjacent to Speedway Motors in Lincoln, NE (see previous page for contact info).

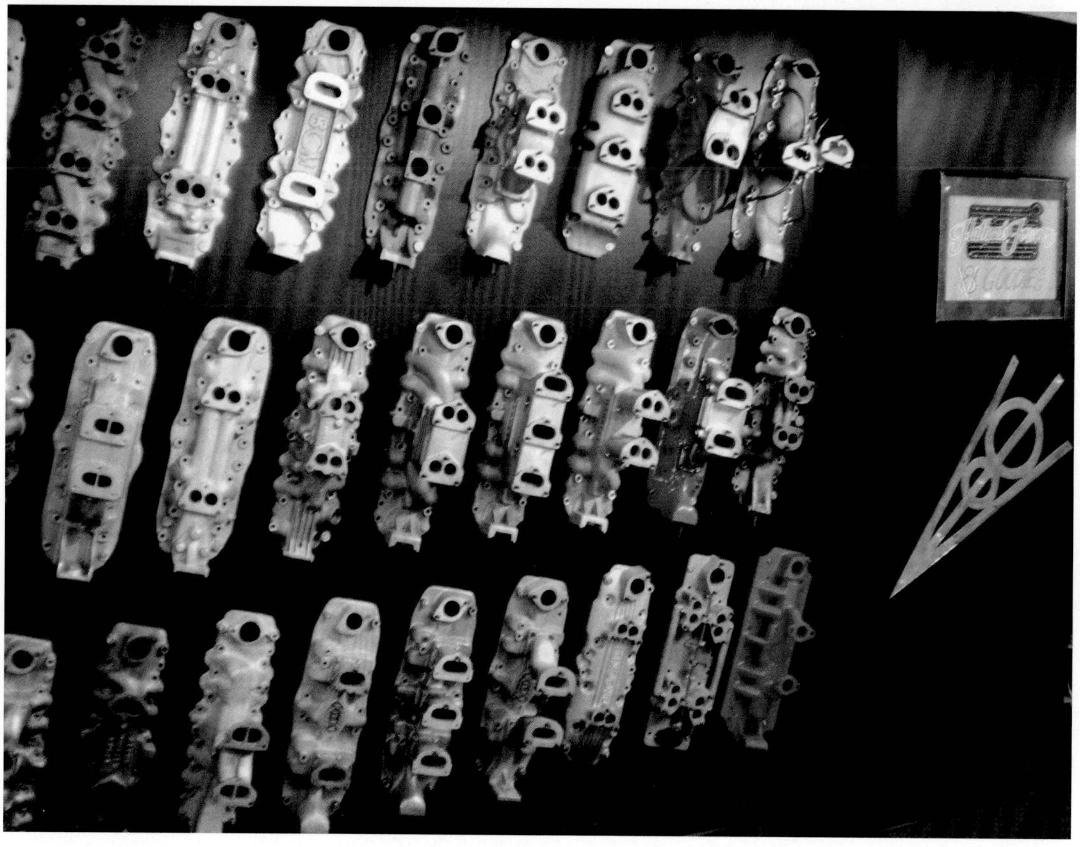

Appendix B

Flathead Cooling

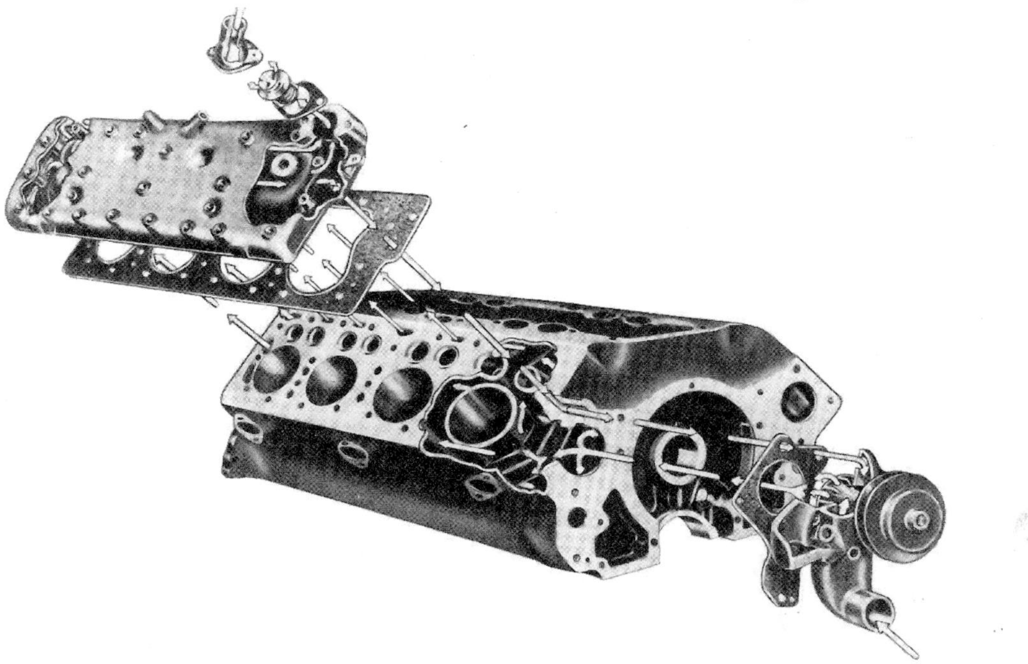

Figure B-1. The Flow of Coolant Through a Mercury *1CM* Engine.

COOLING: GENERAL

As noted in ***Blown Flathead*** and elsewhere, **it is extremely important to start with a clean block**. Assuming the block is clean and the rest of the cooling system is in good shape, there are several other measures we can take to ensure a cool running flathead when we double or even triple the horsepower output. These are listed below.

Check the cooling system to make sure that there are no combustion products getting into the coolant. Any competent radiator shop can check for this. A bad head gasket, a cracked head or a cracked block (real bummer) can result in combustion gases getting into the cooling system, which will guarantee a cooling problem.

Tune the engine. An ignition that is retarded or over-advanced produces less horsepower and more heat. Likewise, a lean air-fuel mixture (jets too small, vacuum leak, etc.) causes the engine to run hot. See Chapters 5 and 6 for tuning tips.

Use a fan. Either an electric or a mechanical (if space allows) fan is a necessity for urban street driving.

Make sure air can flow. Ensure a suitable

grill opening, louvers in the hood and side panels, shrouds and air ducts as appropriate. Some after market radiators have **too many** fins per inch and do not allow sufficient airflow, i.e. the air would rather flow around rather than thru the radiator. Seven fins per inch like the original Ford seems to work well even with four rows of water tubes (four rows deep).

Make sure water can flow. Ensure no kinks or obstructions in the hoses. Lower hoses should have internal springs or be sufficiently stiff to prevent collapse when hot.

Restrictors (e.g. a washer with a 5/8" hole) or thermostats in the water outlets in the heads may enhance cooling by increasing the pumping water pressure in the heads and block, thereby discouraging boiling and hot spots. Unfortunately the use of restrictors can also inhibit cooling and thus fitting them is a trial and error process.

Verify function of the thermostats if used. A .125" hole drilled in the non-movable part of the thermostat helps ensure water movement and proper operation of the thermostat as the water heats up. If a thermostat is used, a 160 degree unit is recommended.

See Figure B-2. This illustration from a Ford service manual shows the proper head gasket orientation for 1949 and later. Note the larger cooling holes at the rear of the gasket to promote coolant flow to the back of the engine and into the rear of the head from the block.

COOLING: LATE HEADS VS. EARLY HEADS

For 1949 and up, Ford redesigned the flow passages in the head gaskets, the block and the heads to promote more uniform cooling of the cylinders and eliminate hot spots. With the later ***8BA*** design, more coolant is encouraged to flow from the water pumps back through the block. The coolant then flows through the enlarged holes at the rear of the block and into the rear of the heads. From there it flows to the front of the heads and back to the radiator through the head outlets, which were relocated in the front of

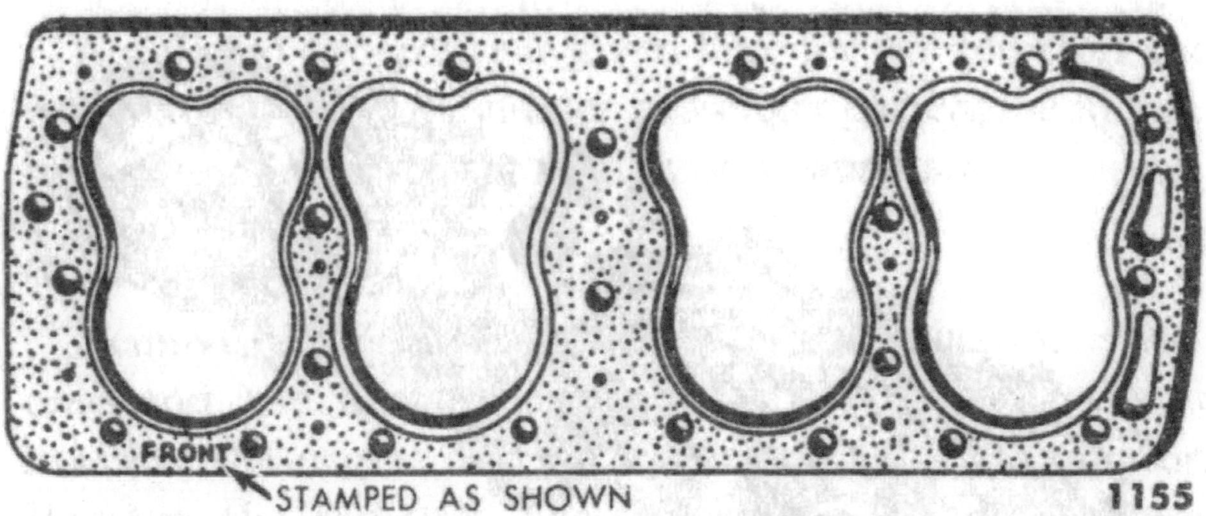

Figure B-2: Illustration from Ford Service Manual Showing Head Gasket and Proper Orientation for 1949 and Later.

the heads. Even though the heads will fit, **do not use the 1948 and earlier heads (or head gaskets) on 1949 and up engines**. This is commonly done for appearance reasons, but cooling capability may be compromised. It is also very important to orient the head gasket properly on the 1949 and up engines as marked on the gasket.

Figure B-3. Flathead Water Pump Impellers. From left to right, we have pre-1938, 1938-1948 (1938-1947 Truck) and 1949 -1953 (1948-1953 Truck) impellers.

COOLING: LATE VS. EARLY WATER PUMPS

The first flathead V-8 water pumps had only 3 or 4 blades or vanes. In the midyears they had 6 vanes. The 1949 and later water pumps flow approximately 25% more coolant than previously, and was accomplished by increasing the size and number of vanes or "paddles" from six to eight and giving them a more efficient curved pumping shape. The additional coolant flow and velocity enhances heat transfer substantially. The additional pumping pressure also increases the coolant pressure in the engine cooling passages thus discouraging the formation of vapor bubbles and hot spots. At the same time, the water pump seals and bearings were upgraded for increased reliability. If pulley location allows, late water pumps may be used on early engines, but the impellers must be trimmed as described in ***Blown Flathead***. The bypass circulation hole at the top of the late pumps ("B" in the Figure B-4) must also be plugged when installing the late pumps on the early engines. The stock early pumps require engine oil for lubrication and cannot be used on later engines. Speedway Motors and others can

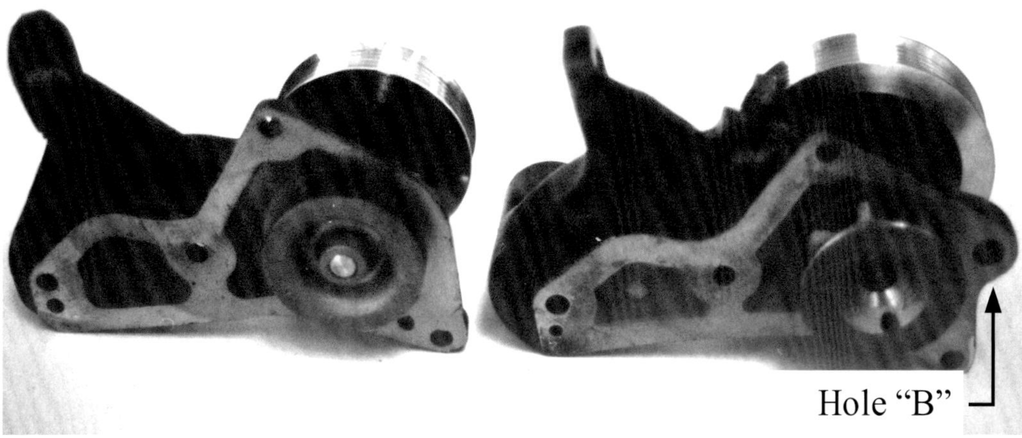

Figure B-4. Engine View of a 1938-1948 Water Pump (left) and a 1948-1953 Truck Water Pump (right).

Figure B-5. Side View of the Water Pumps in Figure B-4.

provide redesigned early-style pumps with self-contained lubrication, better bearings and better seals like the later pumps. Units from Bob Drake also have the late-type high flow impellers. The late high flow impellers may be installed on the early pumps with 5/8" shafts but may require trimming as noted above.

Removing impeller vanes does not make sense except for sustained high speed engines with excess pump capacity (no street flathead engine falls in this category).

COOLING: RADIATOR PRESSURE CAPS

Ford introduced a pressurized cooling system for the flathead in 1941. In 1949, the overall cooling system pressure was increased from four psi to seven psi. Prior to 1949, the radiator pressure cap and the structural strength of the rest of the system had limited the system pressure. The additional system pressure further increases the coolant pressure in the engine over and above the levels achieved by the later, larger water pumps, thus further discouraging the formation of vapor bubbles and hot spots. Of course, the higher pressure also increases the boiling point of the coolant and increases the cavitation (vapor formation at the pump inlet) resistance of the water pumps. It is generally desirable to increase the pressure in the cooling system to 12 psi or more when modern radiators are installed. Caution must be used for the early cars with the large upper radiator tanks like the 1934, because the pressure acting on the large upper tank area can cause catastrophic failure. Also, the stock water pump seals on the pre-1949 engines may be marginal at high pressures. Therefore, the system pressure should be limited to 9 psi or less, even with new construction radiators, if the large tank design is retained or the pre-1949 water pumps are used.

Appendix C

Roadrunner Engineering Flathead Tech Tips

INTRODUCTION

Detailed information on flathead engine preparation and tuning is available in this book and ***Blown Flathead*** available at fine bookstores everywhere. Supplemental tech tips listed below may also be helpful. For questions contact Roadrunner Engineering via the website, www.roadrunnerengineering.com. New tips may also be found on this website.

TECH TIP INDEX

- Exhaust Dividers
- Crankcase Breathers and Positive Ventilation Systems
- Ford Head Bolt Strength
- Installing Stainless Steel and Aluminum Threaded Parts
- Re-torquing Flathead Head Bolts for Permanent Seal
- Ignition Timing For Optimum Power and Economy
- Marking Crankshaft Pulley for Proper Timing
- Retaining Flathead Water Pump Pulleys
- One Piece Front Oil Seals
- Maintaining Gear Lube Levels For Weiand and Other Superchargers
- Modern Transmissions for the Flathead
- High Volume Oil Pump for the Flathead
- Motor Oil for the Flathead
- Oil Filters for the Flathead
- High Flow Air Filters: Pros and Cons
- Electric Fuel Pumps

EXHAUST DIVDERS

Exhaust dividers are typically airfoil shaped cast iron or flat stock pieces that are inserted in the center exhaust ports near the top of the block to prevent intake-exhaust cross contamination between the Siamesed center cylinder exhaust ports. They are usually retained by welding and/or special short head studs that trap and locate the pieces. They do not affect flow numbers appreciably. The only way to tell if they affect power is to dyno test with and without them on the same engine. To my knowledge there is no available test data that supports their use one way or another. Due to the installation effort required and the questionable benefit, I don't believe they are worth the effort on a street engine.

Figure C-1. Cast Iron Exhaust Dividers.

CRANKCASE BREATHERS AND POSITIVE VENTILATION SYSTEMS

All engines need a good breather to vent the crankcase. Even new engines allow leakage of combustion gases ("blow-by"), which end up in the crankcase. If not vented these gases can produce significant crankcase pressure causing major oil leakage out the crankshaft seals. These gases also result in water condensation and acid buildup in the oil, shortening engine life. Recently I replaced the breather cap on the 1936 engine in my '32 with a used, but better-looking cap. My formerly low leak engine began leaking about one quart in a 20-mile drive! The replacement cap was found to be severely restricted even though it had been cleaned and looked fine. A new cap cured the problem. A "road draft" ventilation system was used on later flatheads but a modern positive crankcase ventilation system which routes the blow-by to the intake manifold was never offered on the flathead. If such a system is installed, the fuel delivery curve must be adjusted to handle the additional intake air. **A PCV system venting to the blower intake on a supercharged engine can cause the air-fuel ratio to go lean and initiate detonation by introducing an oil-laden air mixture when boost (and blow-by) are maximum.** The safest approach is to vent the breather to the atmosphere directly or route the vent line above the carburetor inlet (e.g. air cleaner base).

FORD HEAD BOLT STRENGTH

Ford head bolts are a good product! I performed a study where I measured the hardness (correlates with strength) of two brands of commonly available grade 8 head bolts and several stock Ford flathead head bolts. The stock pieces were identified with either a script or block letter "F" and the numbers "1041." The Ford bolts were harder (and therefore stronger) than the grade 8 pieces! NOS or good used Ford head bolts will give good service in any high performance engine.

INSTALLING STAINLESS STEEL AND ALUMINUM THREADED PARTS

An anti-seize compound of some type is recommended for stainless steel or aluminum threads to prevent galling. Also, care should be taken not to over tighten these threads since most are equivalent to grade 2 or less. Permatex thread sealant, P/N 14A, provides anti-seize, thread sealing and some locking capability. Medium strength (blue) thread locking compound provides some anti-seize protection and sealing qualities while providing adequate thread locking where lock washers are not used. Lock washers are not provided in Roadrunner kits.

RE-TORQUING FLATHEAD HEAD BOLTS FOR PERMANENT SEAL

Supercharging results in much higher cylinder pressures than encountered with normal aspiration, and requires a good head-to-block seal. After initial head installation and head bolt torquing, five warm-up, cool-down, and re-torque cycles are recommended before operating the engine at full throttle. Ford recommends three re-torque cycles for stock engines.

IGNITION TIMING FOR OPTIMUM POWER AND ECONOMY

There is an optimum ignition timing for every combination of engine, fuel & operating condition. As blower and cylinder pressure go up, the engine needs less spark advance for peak power, so retarding the ignition is a good thing for power under boost. Retarding the ignition timing, either manually or automatically, is also often desirable to decrease detonation tendencies on supercharged engines, especially when using lower octane fuel. MSD boost retard modules (P/N 5462 or 6462) and others, provide automatic electronic boost retard for most ignitions. These modules are recommended for all supercharged engines. The flathead is a fast burn engine and requires much less advance than most OHV engines. Unfortunately, it is difficult to determine optimum settings without a dyno, so be conservative!

MARKING CRANKSHAFT PULLEY FOR PROPER TIMING

All Roadrunner crankshaft pulleys are provided unmarked. The user must provide the timing marks. After TDC is located (preferably before the heads are installed) using a dial indicator or positive stop, the pulley should be marked appropriately with a scribe or small triangular file. Additional marks should then be spaced 0.00873 x OD of pulley (in inches) per degree. For example, the separation distance for 2 degrees on a 6" pulley is:

2 x 0.00873 x 6 = 0 .1" (approximately).

RETAINING FLATHEAD WATER PUMP PULLEYS

On installations requiring water pump pulleys or hubs to be press fitted, a press force of several hundred pounds is required for proper retention. Manufacturing and or rebuilding tolerances may not provide the desired fit. If the parts present a loose press or no press, additional retention methods should be used. Options include tack welding, installation of setscrews or pins, knurling, or application of adhesives such as Loctite ***620*** or Permatex ***High Strength Sleeve Retainer***.

ONE PIECE FRONT OIL SEALS

The flathead is normally equipped with a rope-type seal at both ends of the crankshaft. These seals generally give satisfactory performance, although there is a certain amount of art involved in installing them properly. Read the instructions for pre-soaking and cutting as several materials are available. Also make sure the mating seal sleeve spiral grove goes the right direction to carry leaking oil back to the pan.

Several suppliers offer a one piece lip seal to replace the front rope seals. The problem is that neither the flathead pan nor the front timing cover are built to properly support the seal and the new seal is generally retained by using a silicone sealant. Also, the original seal sleeve groove is a problem with the lip seal and a smooth sleeve should be installed if the lip seal is used. Not worth the effort in my view.

MAINTAINING GEAR LUBE LEVELS FOR WEIAND AND OTHER SUPERCHARGERS

The blower lubricant level should be approximately halfway in the front sight glass or between the marks for dipstick equipped units. If additional lubricant is required, add 80W-90 or 85-140W gear oil meeting GL-5 specifications through the brass plug on top of the front cover. Do not use synthetic gear oils in these units because the seals are not compatible with these lubes. The rear bearings are sealed and should not require attention.

MODERN TRANSMISSIONS FOR THE FLATHEAD FORD

The stock manual transmissions for the flathead give satisfactory performance under normal driving conditions, but will not survive hard launches reliably, even with stock engines.

Not to worry since there are several options for upgrading the transmission both manual and automatic. Five speed (including overdrive) manual transmissions such as the Borg-Warner ***T-5*** or Tremec work great because they are compact and efficient and several adapters are available. The Ford ***C-4*** automatic (three speed) works well for the same reasons. The ***C-4*** is the only automatic that does not usually require metal cutting to install in an early Ford. An add-on aftermarket unit is required if overdrive is desired. There are also adapters for almost any electronic or automatic overdrive transmission, but these units are large and they often require major surgery to install. They also use significant horsepower just to drive the transmission.

HIGH VOLUME OIL PUMPS FOR THE FLATHEAD FORD

As many flathead owners have observed, the idling oil pressure for a warm flathead is often 10 psi or less. While 10 psi is sufficient at idle (it should rise approximately 10 psi per 1000 rpm) it is not very comforting to those used to the higher numbers from modern engines.

Melling Engine Parts manufactures a high volume oil pump for the flathead designated the ***M15,*** which delivers 25% greater volume capacity than their model ***M19,*** which is the stock equivalent. Peak oil pressure is set by the oil pressure bypass valve (either in the pump or in the engine), however the higher capacity will produce significantly higher oil pressure at low engine speeds. To use the high capacity oil pump in 1948 and earlier engines, a truck oil pickup tube assembly, P/N 8RT-6615 is required. The 1949 and up engines can use their stock oil pickup assemblies. Some bending of the pickup tube may be required to assure adequate clearance (about 3/8" to 1/2") between the oil pickup screen and the oil pan. The oil pressure relief valve in the center valley of the older engines is usually left in place even though the new pumps have a built in relief valve.

MOTOR OIL FOR THE FLATHEAD

In the past we could pretty much assume that if we used motor oil that met the latest American Petroleum Institute (API) specifications, that we were using the best lubrication product and technology for our engines. Then a few years ago, alarming reports surfaced reporting high wear and early failure of engines and engine components particularly those with heavily loaded flat tappets and flat tappet cams. Conflicting studies followed that revealed changes in the parts, the applications, and the oil.

From the available data, it appears that a major contributor to the problems was the reduction of an antifriction oil additive, ZDDP (zinc dialkyldithiophosphate). ZDDP forms a tough film that protects metal parts under extreme pressure conditions. Unfortunately the zinc and phosphorus in this compound can foul modern catalytic converters. For this reason the allowable amount of zinc in most oils was reduced to 870 ppm in 2005, even though many experts feel that the optimum level of ZDDP for wear protection is 1200-1400 ppm. Diesel oils provided an option for a time but they are now also subject to reduced levels of ZDDP. There are additives such as STP that can also boost ZDDP levels, but many of these formulations do not reveal the content, leaving the consumer in the dark as to how much is just right. Too much can foul plugs.

Currently specialty oils are available that maintain a 1200-1400 ppm level of ZDDP. The most common are racing oils, including the best seller, Valvoline ***VR-1***. Racing oils are blended with the primary function being protection of the engine rather than the catalytic converter. The downsides are higher costs and lower detergent levels requiring more frequent oil changes. A good tradeoff I believe. Thus at this time, **I recommend racing oils such as Valvoline *VR-1* for most flathead applications.** At temperatures above freezing, SAE 20W50 meets the requirements of all flathead Fords, stock or modified.

OIL FILTERS FOR THE FLATHEAD

Ford did not offer an oil filter as standard equipment for the flathead until 1940 (earlier it was an option for trucks and other heavy duty applications). All the Ford oil filters for the flathead are "partial flow" or "parallel flow" devices where part of the engine oil is routed to the filter and the major part is routed directly to the engine. These filters can be mounted remotely in any position desired. Although much maligned, these filters work fine and certainly are preferred to no filter. **These filters had a factory fitting with a .060" orifice installed in the inlet line to limit the maximum oil flow to the filter. Make sure this fitting is installed.**

Several conversions for "full flow" or

"almost" full flow filters have been offered commercially or built by owners for years. While these work satisfactorily in most cases, **there can be instant engine failure if a block converted for a full flow oil filter is used without a filter or the wrong filter.** Check the passages in the "boss" at the rear of the block to determine if your block has been converted (one of the passages will be plugged) and use the appropriate filter.

It should be noted that all "full flow" filters are not full flow all the time, and bypass the oil directly to the engine when the oil is cold or the filter is clogged.

HIGH FLOW AIR FILTERS: PROS AND CONS

There are several air filter systems promising more horsepower by allowing more airflow. Most of these filter elements are of the cotton gauss type, which are oiled to enhance dirt capture. The good news is that the filters do flow more air for the same size as a standard pleated paper filter. The bad news is that they also flow more dirt. Using a well known manufacturer's own data for one of these filters vs. the SAE data for a standard pleated paper filter and calculating the difference in filtering for a 302 cubic inch engine traveling 12,000 miles, the difference was over one more pound of dirt ingested by the engine with the high flow filter! This is particularly harmful on tow vehicles which may spend long times at high throttle opening in dusty conditions. There is definitely a trade-off here. You can often maximize airflow by eliminating the filter altogether if engine life is secondary.

ELECTRIC FUEL PUMPS

The stock flathead mechanical fuel pump has a tough job operating in one of the highest and hottest environments in the vehicle. This environment combined with fuel pump aging and pushrod wear results in a flathead trademark - "vapor lock."

To avoid vapor lock, an electric fuel pump mounted as low and as close as practical to the gas tank is recommended for most driver applications. There are many choices for both 6 and 12 volt units, many with built-in pressure regulators and fuel filters. Whether a pump with a built-in regulator or an external regulator is used, the pressure output must be matched to the carburetor requirements. The traditional Stromberg ***97*** inlet pressure should not exceed 3-1/2 psi, for example. A small pressure gage mounted near the carburetor fuel inlet is recommended to monitor fuel pump function and pressure.

If the stock fuel pump and pushrod are not retained, the fuel pump pushrod hole must be plugged to prevent loss of oil pressure. If the engine is not disassembled, a safety-wired, press-in plug is a practical approach.

Figure C-1. Fuel Pump Pushrod Hole Plug Installation. A light press fit is used with a safety wire backup for retention. *Jeff Andreen Photo*

Appendix D

Sir Harry Ricardo and the Ricardo Head

Figure D-1. Sir Harry Ricardo

Harry Ricardo was an English mechanical engineer, born in 1885. Although he died in 1974, the Ricardo Engineering firm he founded in 1915 still exists as a premier engine development company. King George VI knighted Ricardo in 1948 for his enduring contributions to aircraft, marine, automotive, and truck engine systems. Ford and other US auto manufacturers snubbed Ricardo in the 20's and 30's while they were blatantly infringing on his company's patents. In 1955 however, the US Society of Automotive Engineers (SAE) awarded Ricardo the ***Horning Medal***, and he was acclaimed the "high priest of the internal combustion engine." The SAE published his autobiography, ***The Ricardo Story***, in the US, in 1992. He is finally being accorded worldwide recognition for his great work.

His gasoline and diesel engine development contributions were many. He became famous early on in life for developing the

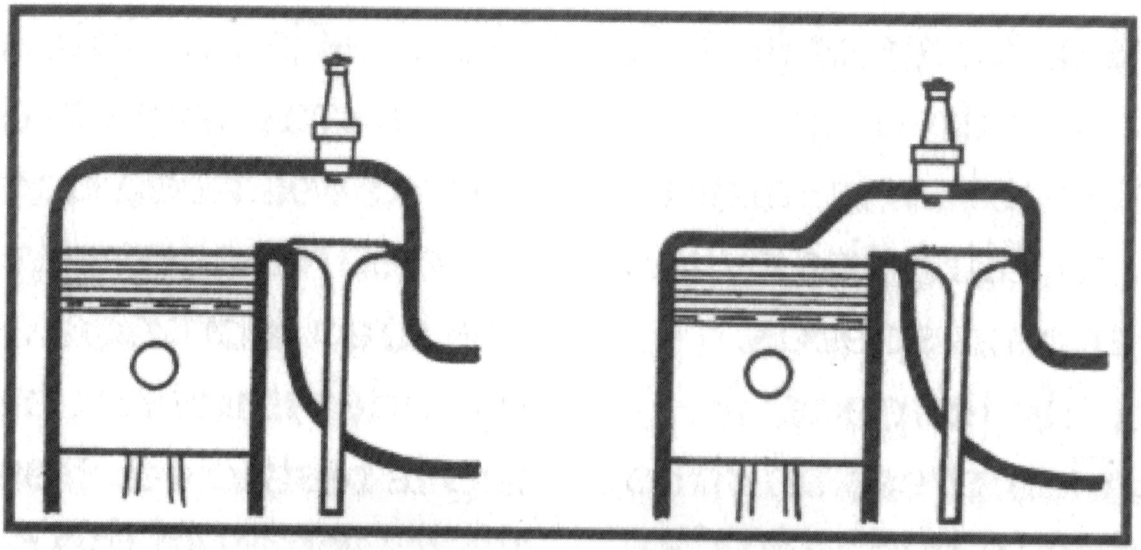

Figure D-2. A conventional internal combustion engine head design in 1919 is shown on the left. The Ricardo turbulent head design is shown on the right. Diagram courtesy of Hot Rod Press.

relationships between engine head design and fuel characteristics on the detonation or "knock" tendencies of engines. His pioneering work in fuels resulted in the "octane" rating system for gasoline in 1931. For early Ford V-8 fans however, his most important development was what he called the "Turbulent Head" design for flathead gasoline engines, which he patented in 1919. His Turbulent Head design improved the power output and fuel economy of the flathead to such an extent that the overhead valve engine did not become dominant for another 30 years. Henry Ford recognized the significance of Sir Harry's invention and incorporated it into the flathead V-8 design in 1932. Ford refused to acknowledge Ricardo's patent or to pay royalties on the use of the patent until he was sued and lost in a landmark decision in 1933. Earlier, Ford had suggested that Ricardo should pay Ford for the "privilege" of incorporating the turbulent head design into a Ford engine! Henry played hardball!

A diagram of the turbulent head design compared with a previous conventional head design is shown above. What Ricardo found was that a large clearance space over the piston remote from the valves encouraged premature ignition or detonation of the air-fuel mixture in that area. The result was poor fuel economy, lost power, and in many cases, engine damage. Conventional thinking at the time suggested making the clearance even larger to reduce the compression ratio and detonation or "knock" tendencies. This worked with extreme clearances but at the expense of fuel efficiency and power. What Sir Harry did was reduce the clearance as shown in the figure to a tenth of an inch or less no matter what the shape of the piston. This relatively small clearance section of the combustion chamber is called the "squish" area. **Most Ford flatheads use heads with a dome shape head over the piston, as shown in Figure 3-14, and thus a domed piston, that matches the shape**

of the head, must also be used. The thin "squish" area allowed the relatively cool piston and head to keep the thin space cool enough to prevent ignition until the flame front incoming from the sparkplug did the job as intended. The thin "squish" area also greatly improved turbulence and mixing of the air-fuel mix. The result was smoother running, higher compression ratios, better fuel economy, lower octane gasoline requirements, and more power. Wow! That's why all flathead V-8 fans have got to be "wild about Harry!"

For more information on the turbulent head and other Ricardo innovations, I highly recommend early (fourth and previous) editions of Ricardo's book, ***The High Speed Internal Combustion Engine,*** published by Blackie and Sons, and his autobiography, ***The Ricardo Story***. Another good book is ***Engines and Enterprise – The Life and Work of Sir Harry Ricardo,*** by John Reynolds, Sutton Publishing, 1999.

SERVICE BULLETIN

SUBJECT NO. S-6000 **PAGE NO. 60**

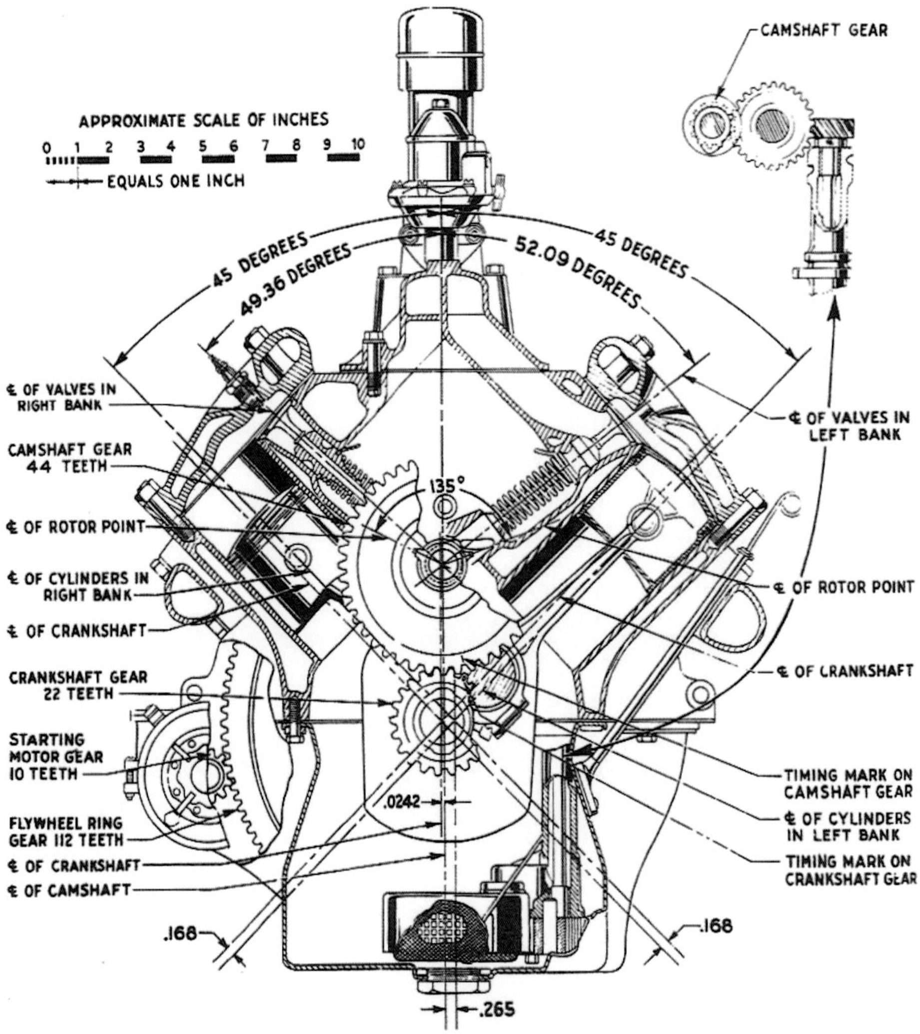

1939 95 H.P. Engine Showing Typical Arrangement

Fig. 2

95 H.P. STARTING 1939

February 15, 1939

Figure D-3. Ford Service Bulletin Showing Flathead Ford Geometry. The figure shows the Ricardo "Turbulent Head" design as discussed in the previous pages. The "Turbulent Head" design is also shown in many of the Chapter 4 figures. Other items of interest on this drawing are that the crankshaft centerline is offset from the cam centerline, as well as from the cylinder bore centerlines. Also, note that the valve centerline angles differ from one side of the engine to the other.

Appendix E

Typical Supercharger Kit Installation Instructions

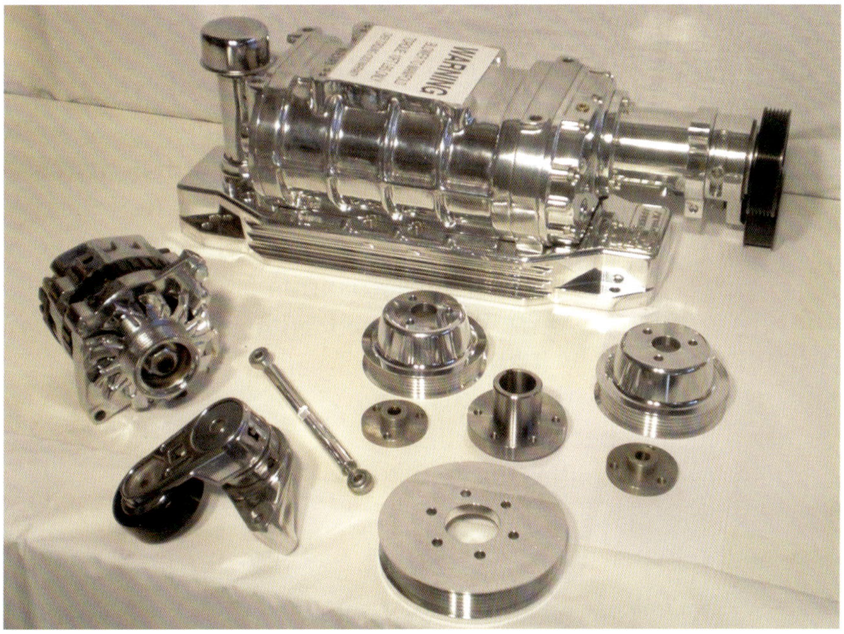

Figure E-1. Typical Parts Included in a Roadrunner Engineering Supercharger Kit.

GENERAL

1. Your supercharger is a precision machine with very tight clearances and tolerances required for proper operation. Cleanliness in all phases of assembly is important. **Failure to follow these instructions can cause major damage to the blower, which is not covered by warranty**.

2. **No lock washers are provided with the Roadrunner kits.** Medium strength (blue) thread locking compound provides some anti-seize protection and sealing qualities while providing adequate thread locking where lock washers would normally be used.

3. The Roadrunner kits for the late flatheads with top mount distributors are designed to use the compact Mallory distributor. The Vertex magnetos and magneto-type distributors also fit. The larger MSD and HEI type distributors will require pulley spac-

ers and/or belt rerouting. Call for details.

4. Roadrunner street kits with a single serpentine belt are designed for Delco 121 series, internally regulated alternators. Polished units are available ready to bolt on from Roadrunner, in a 2-wire, 80-amp configuration (P/N ALT7937-P). See catalog. These alternators are also available locally in satin finish (Lester rebuilder P/N 9737) but must be modified by shortening the long mounting "foot" approximately 0.9" to line up the drive pulley when mounted. Ford 7000 series, Denso and other smaller alternators can also be used when modified as above. **The large GM alternators will not fit.**

5. **Drive belts are not provided with the Roadrunner kits**. The required serpentine drive belt for all kits should be fitted locally because of all the combinations of pulleys and assemblies possible. These belts are commonly and inexpensively available at any auto supply.

6. The following are typical installation instructions for the Roadrunner street supercharger kits for Ford flatheads. Additional instructions may be provided with each kit. See also Chapter 6, Supercharging the Flathead and Appendix C, Tech Tips in ***335 HP Flathead Ford V-8 Performance Handbook***.

PRELIMINARY DISASSEMBLY

1. Disconnect the battery. Drain the cooling system.

2. Remove existing intake manifold, carburetors, generator, fan, etc. Clean gasket surfaces. Clean the existing threaded holes with the proper tap.

3. Remove the existing water pumps. Clean the gasket surfaces. Clean the threaded holes with the proper tap.

4. Remove the existing crankshaft hub/pulley assembly.

WATER PUMP DISASSEMBLY / ASSEMBLY FOR 1938-1953

Note: Original stock 1938-1948 water pumps can not be used with a serpentine drive. New units from Speedway and others have a ball bearing system like the later pumps and can be used. The Roadrunner 2-belt Nostalgia kits can be used with unmodified original pumps and disassembly is not required.

1. Press off the existing pulleys being careful to apply pressing loads to pulley and shaft only. See Figure E-2.

2. Press on new pulley hubs such that hub and shaft outer surfaces are flush as shown in Figure E-3. A press force of several hundred pounds is required for proper retention of the hub on the shaft. Manufacturing and or rebuilding tolerances may not provide the desired fit. If the parts present a loose press, additional retention methods should be used. Options include tack welding, installation of setscrews or pins, knurling, or application of adhesives such as Loctite ***620*** or Permatex ***High Strength Sleeve Retainer***, which are designed for this purpose.

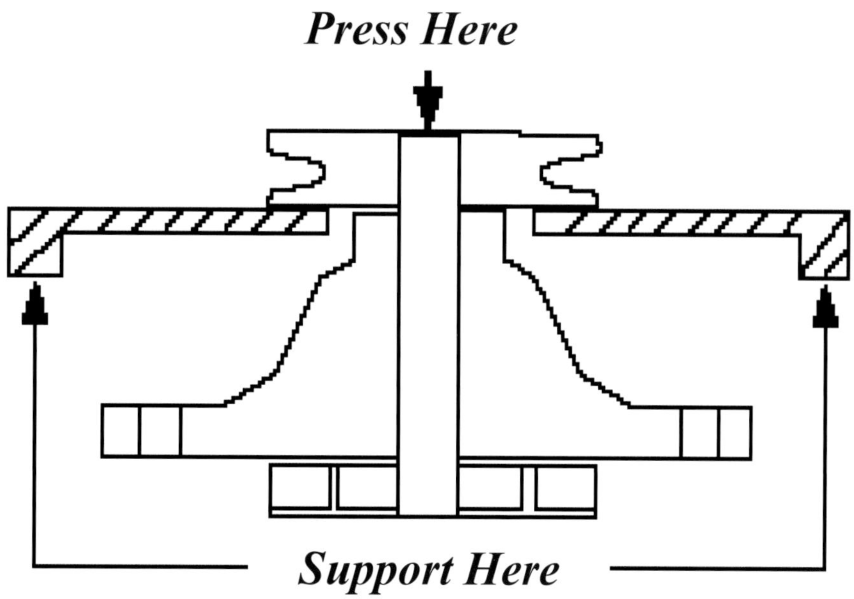

Figure E-2. Removing the Stock Pulley from the Water Pump.

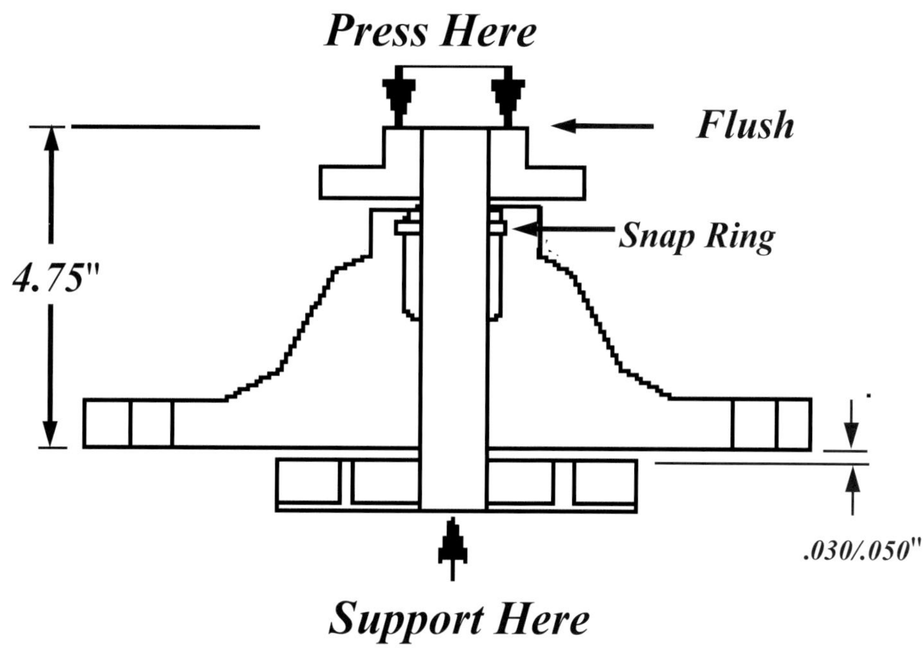

Figure E-3. Pressing the New Pulley Hub on the Water Pump.

ASSEMBLY

1. Install new crankshaft hub on engine. 1938-1948 engines will require a spacer, Ford P/N 8BA-6362.
2. The crankshaft pulley, hub and spacer are shown in Figure E-4. This spacer will already be in place on newer engines. Torque 5/8"-18 retaining bolt to 135 ft-lbs.
3. Install water pumps using thread sealant (Permatex *#14A* or equivalent) on all 3/8"-16 retaining bolts and torque to 15-18 ft-lbs.
4. Install pulleys on water pump and

Figure E-4. Roadrunner Crankshaft Pulley, Hub and Spacer.

crankshaft hubs. Torque 5/16"-24 retaining bolts to 12 ft-lbs.

5. Add timing marks to crankshaft pulley before final installation. The user must provide the timing marks. After TDC is located (preferably before the heads are installed), the pulley should be marked appropriately with a scribe or small triangular file. Additional marks should then be spaced .00873 x OD of pulley (in inches) per degree. For example, the separation distance for 5 degrees on a 5.75" OD pulley is:

   5 x 0.00873 x 5.75 = 0.25" (approx).

6. Plug the fuel pump push rod hole in engine. The mechanical pump and push rod will not be used. **If this hole is left open, oil pressure will be lost.**

7. *Optional. Drill and tap the intake manifold for a boost/vacuum gauge line fitting. A boost gage is desirable for tuning and trouble-shooting. Autometer, VDO and others make suitable gages that include installation hardware. The gage pressure tap is typically an 1/8 inch pipe fitting which can be installed anywhere on the manifold that taps into the intake air plenum (where the intake ports are). A typical pressure tap for a Weiand **142** supercharger is on the top of the manifold, in back of the blower, centered, and approximately 4 -3/4" from the back of the manifold.*

8. *Optional. Install the blower manifold on the engine. Temporarily check port alignment between engine and*

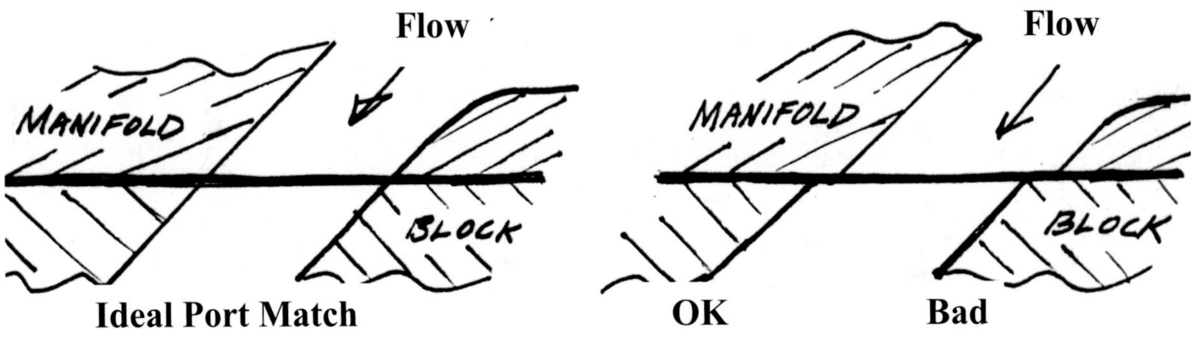

Figure E-5. Port Alignment Between Intake Manifold and Block.

manifold with gasket in place. Match ports and gasket by grinding or cutting as necessary. This is optional but an important step for high performance. It is particularly important that the engine port does not protrude past the blower port. See Figure E-5.

9. Install blower manifold and manifold gasket, then torque 3/8"-16 retaining bolts to 15-18 ft-lbs. 1949 and later engines will not have as many mounting bolts as earlier engines (13 versus 17). Competition or high boost (>8 psi) engines should consider using the maximum number of bolts to enhance sealing. In this case, later blocks can be drilled and tapped appropriately.

10. Install blower assembly and gasket. Coat the gasket with talcum or baby powder to prevent sticking to manifold or supercharger. **Do not use any type of gasket sealant as this will void your warranty.** Make sure you have removed any tape residue with solvent after removing tape from bottom of the blower and take care not to let any foreign matter into the rotor housing. **Torque the 7/16"-14 retaining bolts to 8-10 ft-lbs using a criss-cross pattern.** While you are tightening the supercharger retaining bolts, turn the supercharger drive pulley by hand to make sure there is no bind. After tightening, remove the tape cover from the top of the supercharger and remove any residue.

11. Using a .010" feeler gauge, check the rotor to case clearance all along both of the upper edges of the case from front to back, inserting the gauge no more than ½" past the edges. If there is any bind against the feeler gauges, loosen the retaining bolts and retorque.

12. Check lubricant levels in blower. The blower lubricant level should be approximately halfway in the front sight glass or between the marks for dipstick equipped units. If additional lubricant is required, add 80W-90 or 85-140W gear oil meeting GL-5 specifications through the brass plug on top of the front cover. Do not use synthetic gear oils in these units because the seals are not compatible with these lubes. The rear bearings are sealed and should require no attention for the life of the unit.

13. Install custom Delco ***121*** series alternator, Roadrunner P/N ALT7937-P, to mount on the supercharger nose. This is an internally regulated, 2-wire alternator. The alternator "Bat" post is connected to the battery (usually connected at the starter solenoid). The alternator "F" terminal is connected to the ignition switch.

14. Install tensioner rod to alternator and timing cover. 1948 and earlier engines will require removal of the vacuum brake on the distributor and plugging of the hole. The vacuum brake will not be used.

15. If used, install optional automatic belt tensioner, Roadrunner P/N 4013 (recommended) with the single serpentine street kits. See Figure E-7. The belt length for these kits should be chosen such that the automatic tensioner arm is approximately in the middle of the marked operating range. See Figure E-6. Note the **Nostalgia**

Figure E-6. Automatic Tensioner With Adjustment Marks. The single mark on the right should ideally match up with the middle mark on the left.

and **Competition** kits have a different automatic belt tensioner as standard equipment. The belt length for these kits should be chosen such that the tension arm/idler pulley is approximately at the 7:30 (clock) position when viewed from the front.

16. On street kits **without** an automatic tensioner, adjust the tensioner rod to get sufficient belt tension. A jam nut is provided to lock the rod position. The tension is sufficient when the alternator pulley can **not** be rotated easily by hand. Too little tension will allow the belt to slip under high boost. The symptoms may include squealing, low boost, and belt dust on the firewall. Too much tension can damage the blower, water pumps, alternator and even the engine. Recheck belt tension often initially and periodically thereafter. The variable position of the alternator bracket and the tensioner rod adjustment allow a belt to be fitted and tensioned easily.

17. Install serpentine belt. See Figure E-7 for belt routing. As an example, if you have a street kit with the automatic tensioner for the 1949-1953 flatheads with a 5.63" drive pulley and a 3.75" blower pulley, you will use a 6 rib, "***K***" series belt, 79.5 inches long. The Goodyear P/N is 6PK795 (4060795).

18. Install the oil breather/oil filler assembly at the rear of the blower manifold. A stock unit may be used or an accessory item such as Roadrunner P/N FB-44 looks great. The road draft breather system on the late flatheads is **not** used. The early flatheads have an additional breather at the front of the pan. A similar breather such as those made by Moroso can be added to the late flathead pans if needed

19. A positive crankcase ventilation system venting to the blower inlet is **not recommended** for any carbureted supercharged engine, because it can cause the air-fuel mixture to go lean and initiate detonation by introducing an oil-laden air mixture when boost (and blow-by) are maximum. The safest approach is to vent the breather to atmosphere directly or route above the carburetor inlet (e.g. air cleaner base).

20. Install the carburetor on the blower using the proper studs and nuts, and an appropriate gasket/spacer/adapter as necessary for the carburetor. **Cleanliness is important here**. The Weiand blowers are made to accept almost any four-barrel carburetor including the ***Quadrajet*** style with large primaries. A gasket/spacer assembly which covers the blower intake properly and prevents parts interference must be used. Square-bore carburetors

such as the Holley ***4150*** and ***4160*** series will require a Holley carburetor adapter plate (1/16" thick steel) and gasket set, P/N 9006. A thicker (1/4") adapter such as the Fel-Pro P/N 60441 (13-1666) may be required for some applications for throttle linkage clearance.

21. A 385-570 cfm 4-barrel carburetor with vacuum operated secondaries will give good results in most applications. Make sure the carburetor throttle blades and linkage operate smoothly without binding. Full throttle is essential for maximum performance. **A jammed throttle is extremely dangerous**.

22. Install a high capacity air cleaner (essential). The air cleaner top should be at least 1" above the carburetor choke tower.

23. The Roadrunner street kits for the late flatheads with **top mounted** distributors are designed to use the compact Mallory distributors. The Mallory is available in point or electronic versions and can be used with electronic boost retard modules such as the MSD ***6BTM*** (recommended) and others. The new Vertex magnetos and mag-type distributors look good and also fit. The larger MSD and HEI type distributors may be used, but will require pulley spacers and/or belt rerouting. Call for details.

24. See Chapter 6 of ***335 HP Flathead Ford V-8 Performance Handbook*** for starting points for fuel system, ignition system, and tuneup tips. There is an optimum ignition timing for every combination of engine, fuel and operating condition. As blower and cylinder pressure go up, the engine needs less ignition advance for peak power, so retarding the ignition is necessary for peak power. Retarding the ignition timing, either manually or automatically, is often desirable to decrease detonation tendencies when using lower octane fuel. MSD can provide automatic electronic boost retard modules for any ignition (P/N's 5462 & 6462). Unfortunately, it is difficult to determine optimum settings without a dyno, so be conservative!

25. Connect carburetor linkage and fuel lines. Install electric fan. Refill cooling system and reconnect battery. Start engine, checking for leaks. Check and adjust idle speed, timing and belt tension if required.

26. Supercharging results in much higher cylinder pressures than encountered with normal aspiration, requiring a good head-to-block seal. After initial head installation and head bolt torquing, five warm-up, cool-down, and re-torque cycles are recommended before operating the engine at full throttle.

27. Go forth and kick those overheads around!

Figure E-7. Single Serpentine Belt Routing With An Automatic Tensioner. The proper position of the spring-loaded tensioner arm is as shown above and in Figure E-6. The manual tensioner threaded adjustment rod is retained to properly locate and support the alternator. The belt routing with the manual tensioner only is almost identical to that in the picture except that the belt goes directly from the passenger side water pump pulley to the crankshaft pulley. The manual tensioner gives adequate belt wrap for blower boost levels up to about 6 psi.